# WATER AND LIFE

## BY DR KALTOUM BELHASSAN

# CONTENTS

# LIST OF FIGURES

# INTRODUCTION

In the words of famed oceanographer Jacques Cousteau, "the earth is a water planet." (Zronik 2007). Water makes life, as we know it, possible. There is no life without water. Water and life are inseparable. No known living thing can function without water, and there is life wherever there is water on Earth (Rothschild and Mancinelli 2001). Every drop of water cycles continuously through air, land, and sea to be used by someone (or something) else "downstream."

Water is a chemical compound with the formula $H_2O$. A water molecule contains one oxygen and two hydrogen atoms connected by covalent bonds. Water is a liquid at standard ambient temperature and pressure, but it often coexists on Earth with its solid state (ice) and gaseous state (water vapor or steam). Water also exists in a liquid crystal state near hydrophilic surfaces (Henniker 1949).

Water covers about 70 percent of the planet, giving it the unique ability to foster and sustain life. Nevertheless, only 2.5 percent of that water is freshwater, and only a fraction of 1 percent supports all life on land.

Water is the most precious natural resource. It is not just a drink. It is necessary for growing crops, cooking, washing, and more. A growing human population and the expansion of economic activities that comes with it intensify the pressure on water resources. There is a water crisis today. But the crisis is not about having too little water to satisfy our needs. It is a crisis of managing water so badly that billions of people—and the environment—suffer badly (Cosgrove and Rijsberman 2000).

This book aims to show the importance of water for living things and to increase the awareness that our freshwater resources are limited and need to be protected, in terms of both quantity and quality.

The work is divided into four parts.

# PART I: THE NATURAL HISTORY OF WATER

- ✓ **CHAPTER 1** offers a general introduction about water in the universe.

- ✓ **CHAPTER 2** highlights the distribution of water on Earth.

- ✓ **CHAPTER 3** discusses different forms of water.

- ✓ **CHAPTER 4** discusses the main chemical and physical properties of water.

- ✓ **CHAPTER 5** provides an overview of the circulation of water.

# PART II: WORK DONE BY WATER

- ✓ **CHAPTER 6** describes how rock is weathered (broken down) by water.

- ✓ **CHAPTER 7** is devoted to erosion of the land surface.

- ✓ **CHAPTER 8** describes the deposition of sediments.

# PART III: WATER SUPPLY ENGINEERING

- ✓ **CHAPTER 9** provides basic knowledge about the sources of the water supply.

- ✓ **CHAPTER 10** is about water quality and how people can clean water and make it safe for consumption.

- ✓ **CHAPTER 11** highlights some of the chemistry, biology, and physics involved in the treatment of water.

- ✓ **CHAPTER 12** discusses components of the water system and common methods of water distribution.

# PART IV: WATER USE

- ✓ **CHAPTER 13** is about how water affects life and how primordial it is for all.

- ✓ **CHAPTER 14** considers the effects of water on human civilization.

- ✓ **CHAPTER 15** looks at managing water scarcity with a water policy to protect the lives of all living things.

# Part I:

## THE NATURAL HISTORY OF WATER

# WATER IN THE UNIVERSE

## 1. INTRODUCTION

The universe is conducive to life that requires water, energy, and carbon; an environmental disaster that removes water dooms life. Water is the habitat in which life first emerged and with which all of it still thrives. This chapter aims to give some understanding of the relationship between water and the universe's habitable zones.

## 2. WATER IN THE UNIVERSE

Much of the universe's water is a by-product of star formation. When stars are born, a strong wind of gas and dust flows out of star nurseries. It contains clouds of hydrogen (H) and oxygen (O) that react together to form water ($H_2O$). When this outflow of material eventually impacts the surrounding gas, the shock waves that are created compress and heat the gas. When Earth formed around 4.54 billion years ago through the attraction of many wet dust particles, water was trapped within it. Much of the water on Earth eventually made its way to the surface, giving our planet its relatively finite amount of water at formation.

Water may exist in a variety of places in the universe, including on our moon, Mars, Jupiter's moons, and in comets, because its components, hydrogen and oxygen, are among the most abundant elements in the universe.

Researchers say that water has been prevalent in the universe for nearly its entire existence. The amount of water estimated to be in the quasar is at least a hundred thousand times the mass of the sun, equivalent to 34 billion times the mass of Earth (Taylor 2011).

## ■ 2.1. WATER VAPOR

Water vapor or aqueous vapor is the gas phase of water. It is one state of water within the hydrosphere. Water vapor is produced by the evaporation or boiling of liquid water or from the sublimation of ice.

**Earth's atmosphere**. Water vapor is the dominant greenhouse gas on Earth, the most important gaseous source of infrared opacity in the atmosphere (Held and Soden 2000). Earth's atmosphere contains 0.40 percent water vapor (Cain 2009).

**Mercury's atmosphere** contains water vapor (Rodríguez et al. 2004). Mercury's atmospheric water vapor is around 0.03 percent (Cain 2012).

**Venus's atmosphere.** Astronomers have detected that the atmosphere of Venus contains about 0.002 percent water vapor (Cain 2009). Most of it is in the cloud tops of Venus (Cottini et al. 2011).

**Jupiter's atmosphere**. Scientists were astounded to find water vapor in the freezing atmosphere of Jupiter (Radford 1998).

**Mars's atmosphere.** Observations of Mars in late 1992 and during its early 1993 opposition revealed visible clouds of water vapor in the upper atmosphere (Parker and Berry 1993).

**Atmospheres of Saturn, Uranus, and Neptune.** The ESA's Infrared Space Observatory (ISO) in 1997 measured the concentration of water vapor in the atmospheres of Saturn, Uranus, and Neptune (Ron 1997).

**Enceladus (one of Saturn's moons)**. Cassini's Ultraviolet Imaging Spectrograph (UVIS) observed stellar occultations on two flybys of Enceladus and confirmed the existence, composition, and regionally confined nature of a water vapor plume in the moon's south polar region. This plume provides an adequate amount of water to resupply losses from Saturn's E ring and to be the dominant source of the neutral hydroxide (OH) and atomic oxygen that fill the Saturnian system (Hansen et al. 2006). In addition, the combined observations of Enceladus by The Cassini Plasma Spectrometer (CAPS), Ion and Neutral Mass Spectrometer (INMS), and other instruments detected water vapor geysers. Molecular nitrogen ($N_2$), carbon dioxide ($CO_2$), methane ($CH_4$), propane ($C_3H_8$), acetylene ($C_2H_2$), and several other chemical species, together with all of the decomposition products of water were present in the water vapor geysers (Dennis et al. 2007).

## 2.2. WATER AS LIQUID

Liquid water is $H_2O$'s fluid state, one of the three principal states of water. The liquid form of water is only known to occur on Earth, Enceladus, and Europa.

**Earth**. Liquid water on Earth constantly evaporates into water vapor (clouds), and then condenses back into liquid water (rain) (Gehl 2013). Earth should have been a frozen wasteland, but all geologic signs point to a young planet awash in liquid water, allowing the first life-forms to emerge (Wayman 2013).

**Enceladus**. According to NASA's Cassini spacecraft, this moon orbiting Saturn seems to have liquid water under its icy surface, and organic material is spewing from cracks in its surface (Ledford 2012).

**Europa**. The sixth-closest moon of the planet Jupiter seems to have liquid water. A saltwater ocean has been inferred below its surface, deep enough to cover the whole moon. It contains more liquid water than all of Earth's oceans combined (NASA 2011).

## 2.3. WATER AS ICE

Water becomes solid (that is, ice) if its temperature is lowered below 273 degrees Kelvin (equal to 0°C and 32°F). Water ice is present on Earth, the moon, Europa, etc.

**Earth**. Ice is found on Earth mainly as ice sheets. As glaciers have advanced and retreated, they have sculpted mountain valleys and scoured the land. Water with properties similar to that in our oceans has been discovered in a comet for the first time. This supports the theory that water first arrived on Earth when comets smashed onto our planet's surface. The water in a comet called Hartley 2 was found to have the same chemical composition as our oceans (Fyall 2011).

**The moon**. Since the 1960s, scientists have conjectured that water ice could survive in cold, permanently shadowed craters at the moon's north and south poles. Water that has fallen on the moon (because of comets crashing into it) and has managed to collect in these craters has apparently remained frozen for thousands, if not millions, of years (Steele 2001).

**Europa and Ganymede**. These moons of Jupiter have been found to be covered by extensive areas of ice. Many scientists think that beneath these ice sheets are large areas of liquid water (Steele, 2001).

**Mars**. Scientists have long believed that the poles of Mars contain frozen water (Eddy 2004).

**Titan.** This moon is the largest moon of Saturn. It is primarily composed of water ice and rocky material. Its icy bedrock lies extensively exposed on its surface (Griffith et al. 2003).

**Enceladus**. Only three hundred miles wide, this moon was expected to be nothing more than a frozen chunk of ice and rock. Instead, NASA's Cassini spacecraft spotted eruptions of water vapor and ice particles (Chang 2006). Scientists now have a better sense of where they might originate; they believe they come from underground reservoirs of liquid water, as opposed to the sublimation of water ice (Dornheim 2006).

**Pluto and Charon**. Once thought to be a singular system in an odd orbit at the edge of the solar system, these objects are now known as members of a vast population of icy bodies beyond Neptune (Brown 2002).

## 3. WATER AND THE HABITABLE ZONE

Planets' suitability for life depends on the abundance of certain volatile compounds (especially water) and on their climates. Only planets within the liquid-water habitable zone (HZ) can support life on their surfaces (Kasting and Catling 2003). So, there's no place like Earth. Defining a "habitable zone" in the cosmos seems simple. The existence of liquid water on Earth (and to a lesser extent, its gaseous and solid forms) is vital to the existence of life, as we know it. Earth is located in the habitable zone of the solar system; if it were slightly closer to or farther from the sun by even 5 percent (about eight million kilometers), conditions that allow the three forms to be present simultaneously would be far less likely (Ehlers and Krafft 2001).

In addition, if a planet has no atmosphere, all of the sunlight that strikes it reaches the surface. Thus, our atmosphere is essential for trapping some of the sun's energy, warming Earth enough to support life. This phenomenon is known as the "greenhouse effect." If Earth were smaller, a thinner atmosphere would allow temperature extremes, thus preventing the accumulation of water except in polar ice caps (as on Mars). Conversely, if a planet has a thick, cloud-filled atmosphere, part of the sunlight is reflected into space before it even reaches the planet's surface.

All organisms and their inorganic surroundings on Earth are closely integrated to form a self-regulating, complex system (the biosphere, the atmosphere, the hydrosphere, and the pedosphere) that regulates Earth's surface temperature, atmosphere composition, and ocean salinity. These processes at the Earth's surface are essential to maintain conditions for life on the planet.

Water is the only substance on our planet that naturally occurs in the three physical states (solid, liquid, and gaseous), which directly depend on the atmospheric pressure. If a planet is sufficiently massive, the water on it may be solid even at high temperatures because of the high pressure caused by gravity, as has been observed on exoplanets[1] (Aguilar 2009).

## 4. CONCLUSION

We believe that water exists in abundance in the universe because its components, hydrogen and oxygen, are among the most abundant elements. Conditions on Earth are the best example of a habitable zone. One thing that makes our planet special is the presence of liquid water, which is fundamental for all life; without it, every living thing would die. But what is the distribution of water on Earth? The next chapter answers this question in detail.

---

1  A planet that orbits a star outside of the solar system.

# WATER ON EARTH

## 1. INTRODUCTION

Life is found only on Earth. Life requires water, sun, moderate temperatures, and certain chemical elements, such as oxygen, carbon, and nitrogen. No other planet in our solar system has the same conditions for life as Earth does.

The portion of Earth that is composed of water is called the hydrosphere. "Hydrosphere" derives from the Greek words for "water" and "ball" (Lerner and Wilmoth Lerner 2008). It includes the oceans, seas, lakes, ponds, rivers, streams, and groundwater. The hydrosphere covers about 70 percent of the surface of Earth and is home to plants, animals, and humans. Earth's approximate water volume is about 1.4 billion km$^3$ (Gleick 1993).

When you understand the distribution of Earth's water, you'll understand how important water is for life.

## 2. SALT WATER

Salt water, as in oceans, makes up 97.5 percent of all water on Earth (Figure 1), occupying a volume of 1,365,000,000 km$^3$.

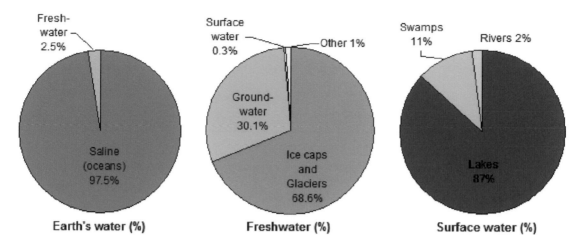

Figure 1. Distribution of Earth's Water

The ocean is the cornerstone of our life-support system. Without it, Earth would be as inhospitable as Mars is. The ocean drives climate and weather, shapes planetary chemistry, generates more than 70 percent of the oxygen in the atmosphere, absorbs carbon dioxide, and replenishes our freshwater through clouds (Earle 2003). The average salinity of Earth's oceans is about thirty-five grams of salt per kilogram of seawater or 3.5 percent (Kennish 2001). Figure 2 shows the South Pacific Ocean from Noumea, New Caledonia.

The ocean is home to most life on Earth. By continuously processing chemicals, water, and minerals, its living creatures shape the way the world works. Yet there is a widespread view that fish and other marine life are limitless commodities.

Other salt water contains 1 percent of the Earth's water volume: saline groundwater (0.93 percent, occupying 13,020,000 km$^3$) and saline lakes (0.07 percent, occupying 980,000 km$^3$).

## 3. FRESHWATER

Freshwater refers to the water in lakes, ponds, streams, and any other body of water besides the sea. Freshwater comes from condensation of the atmospheric water vapor on a surface maintained below the dew point through radiated heat loss to the night sky.

Water—precisely, freshwater—is essential for life. Plants and animals living in freshwater are usually unable to live in salt water, because their bodies are adapted to a low salt content (Figures 3 and 4). Freshwater environments are less extensive than sea environments. The volume of freshwater resources is ~35 million km$^3$, or about 2.5 percent of the total water volume (Figure 1).

Nevertheless, freshwater environments are important centers of biodiversity. This is especially so in dry environments, like deserts, where isolated ponds and streams provide havens for plants and animals.

The total usable freshwater supply for ecosystems and humans is ~200,000 km$^3$ of water, which is < 1 percent of all freshwater resources, and only 0.01 percent of all the water on Earth (Gleick 1993, Shiklomanov 1999).

Figure 2. The South Pacific Ocean from Noumea, New Caledonia

Figure 3.
Plitvice Lake
National
Park, Croatia

Figure 4. Two Swans on Windermere Lake, England, United Kingdom

## 3.1. ICE CAPS AND GLACIERS

Glacial ice is freshwater (not salt water); glaciers contain the largest reservoir of freshwater on Earth. Glaciers and ice caps contain about 24 million $km^3$ or 68.6 percent of the freshwater on Earth (Figure 1). They cover about 10 percent of all land. They are concentrated in Greenland and Antarctica.

Unfortunately, most of these resources are located far from human habitation and are not readily accessible for human use. According to the United States Geological Survey (USGS), 96 percent of the world's frozen freshwater is at the South and North Poles, with the remaining 4 percent spread over 550,000 $km^2$ of glaciers and mountainous ice caps measuring about 180,000 $km^3$ (Untersteiner 1975, UNEP 1992, WGMS 1998).

## 3.2. GROUNDWATER

Groundwater is by far the most abundant and readily available source of freshwater, followed by lakes and rivers. Groundwater comprises about 10,535,000 $km^3$, or 30.1 percent of the freshwater on Earth (Figure 1). This constitutes about 97 percent of all the freshwater that is potentially available for human use (Boswinkel 2000). It is our main source of water for everything from drinking to washing to nurturing the food we eat. About 1.5 billion people depend upon groundwater for their drinking water supplies (WRI et al. 1998). The amount of groundwater withdrawn annually is roughly estimated at 600-700 $km^3$, representing about 20 percent of global water withdrawals (WMO 1997).

## 3.3. SURFACE WATER

Surface water comprises an estimated 105,000 $km^3$, or ~0.3 percent, of the world's freshwater (Figure 1).

**Lakes** contain about 91,455 $km^3$, or 87 percent, of the total surface water. Most freshwater lakes are located at high altitudes, with nearly 50 percent of the world's lakes located in Canada alone. Many lakes, especially those in arid regions, become salty through evaporation, which concentrates the inflowing salts.

**Rivers** contain about 2 percent of the total surface water (Figure 1). Rivers form a hydrologic mosaic, with an estimated 263 international river basins covering 45.3 percent (231,059,898 km²) of the Earth's land surface, excluding Antarctica (UNEP, Oregon State University et al. in preparation). The total volume of water in the world's rivers is estimated at 2,115 km³ (Groombridge and Jenkins 1998). Figure 5 shows Crimea Mountain River, Protected areas of the Crimea Grand Canyon, Crimea.

**Swamps** contain about 11,511.50 km³, or 11 percent, of Earth's total surface water (Figure 1).

## ■ 3.4. OTHER FRESHWATER

Ice and snow, soil moisture, atmospheric water, and biological water are the other places where freshwater is found. They make up about 350,000 km³, or 1 percent, of the freshwater on Earth (Figure 1).

**Ice and snow.** Many boreal continental water bodies have seasonal ice cover. Ice and snow contain about 0.95 percent of the freshwater on Earth.

Ice forms on rivers and lakes in response to seasonal cooling. Frozen lakes and rivers comprise a tiny additional fraction of freshwater. Water is removed from frozen rivers and lakes through the hydrologic cycle.

Snow is more sensitive to elevated temperature than massive ice caps and glaciers are. Accordingly, although snow falls in many temperate regions, it usually does not stay on the ground for any appreciable length of time. In some regions at high altitude, snowfields act as natural reservoirs or supply system. Snowfields store precipitation from the cool season, when most precipitation falls, in the form of snowpacks. In the warm season, most (or all) snowpacks melt and release water into rivers. Snowmelt can cause flooding for many people around the world, mainly during the spring.

Soil is a collection of natural bodies occupying a portion of Earth's surface. **Soil moisture**, the water content in soil, is measured by mass or volume. Soil moisture contains about 0.003 percent of the freshwater on Earth.

Figure 5. Crimea Mountain River, Protected areas of the Crimea Grand Canyon, Crimea

Movement of water into soil is called infiltration, and the downward movement of water within the soil is called percolation. Pore space in soil is the conduit that allows water to infiltrate and percolate and provides a major reservoir for water within a catchment. The ability of soil moisture to affect precipitation (P) can be dissected into the ability of soil moisture to affect evapotranspiration (ETR) and the ability of ETR to affect precipitation (Jiangfeng and Paul 2012). Soil moisture levels rise when there is sufficient rainfall to exceed losses to evapotranspiration and drainage to streams and groundwater. Soil moisture levels are depleted during the summer when evapotranspiration rates are high.

Soil water supports plants. Soil water dissolves salts and makes up the soil solution, which is important as a medium for supplying nutrients to growing plants (Figures 6 and 7). If the moisture content of a soil is optimum for plant growth, plants can readily absorb soil water. Not all water present in the soil profile is available for the use of plants. Much of it remains in the soil as a thin film.

**Biological water** contains about 0.046 percent of the freshwater on Earth. Water is essential to life because it is a major component of cells, typically forming between 70 percent and 95 percent of cell mass, and it provides an environment for aquatic organisms. Almost all proteins and nucleic acids are inactive in the absence of water, and hydration determines their structural stability, flexibility, and function (Chaplin 2006, Pal and Zewail 2004, Levy and Onuchic 2006, Rasaiah et al. 2008, Bagchi 2005, Mattos 2002, Frauenfelder et al. 2009, Zhong 2009). Specifically for proteins, the dynamics of water-protein interactions govern various activities, including the facilitation of protein folding, maintenance of structural integrity, mediation of molecular recognition, and acceleration of enzymatic catalysis (Zhong et al. 2011).

Figure 6.
Soil water,
Cross section
view of a
Plant with
Roots

Figure 7.
Nutmeg
Forest park
in Jeju Island,
Bijarim,
Korea

20

**Atmospheric water.** The atmosphere isn't just made up of air. It also contains water vapor evaporated from the oceans, lakes, rivers, and standing water. Although it is invisible and has no smell, vapor flows around the atmosphere ready to do its important part in creating storms and contributing to life on Earth. There is a considerable amount of water vapor in the atmosphere, though it is less than 0.001 percent of all the water on Earth. It occupies a volume of 12,700 km$^3$ (Lerner and Wilmoth Lerner 2008). However, this tiny amount of water in the air is extremely important to our climate. The greater the amount of water vapor in the air close to the ground, the more uncomfortable it feels on a warm day (Paul 1998).

## 4. CONCLUSION

Seventy percent of Earth is covered by water. The total amount of water in the world is approximately 1.4 billion km$^3$, of which 97.5 percent is salt water, occupying 1,351,000,000 km$^3$. Freshwater lakes and rivers, ice and snow, and underground aquifers hold 2.5 percent of the world's water. There is 35 million km$^3$ of freshwater on Earth. Nearly 70 percent of that freshwater, or 24.34 million km$^3$, is in the form of glacial ice or snow. Groundwater and soil moisture account for 10.55 million km$^3$ (30.1 percent). Only 0.3 percent of it is in freshwater lakes, rivers, and swamps. Of this surface freshwater, 87 percent is found in lakes. Freshwater lakes hold about 0.1 million km$^3$. Rivers, the most visible form of freshwater, account for 0.002 million km$^3$.

# FORMS OF WATER

## 1. INTRODUCTION

Water is indispensable to life. It was not until water became a feature of the planet's surface that life became possible on Earth. Like many substances, water can take numerous forms that are broadly categorized by phase of matter: liquid, solid, and gaseous states.

This chapter aims to give readers the background information that they need about the forms of water.

## 2. FORMS OF WATER

Water makes up three-quarters of the Earth's surface. While we know more about Earth's land areas, water is a vital part of the planet. As we have seen, water on Earth can be either salt water, as found in oceans and some seas, or freshwater, as found in rivers, lakes, springs, and groundwater.

### ■ 2.1. OCEANS

Ocean water is salt water, which is not for drinking. There are four oceans. The largest is the Pacific Ocean (covering about one-third of the world). The next largest is the Atlantic, then the Indian, and finally, the Arctic Ocean.

The oceans together have an area of about 131 million square miles (Figure 8). Connected or marginal to the main ocean basins are various shallow seas, such as the Mediterranean Sea, the Gulf of Mexico, and the South China Sea.

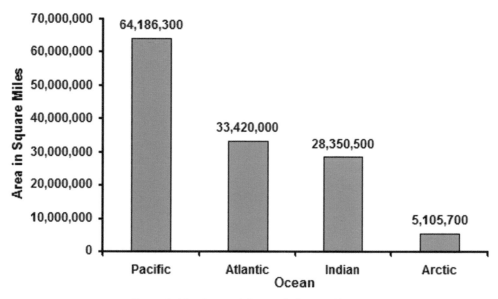

Figure 8. The Areas of Oceans in Square Miles

The deepest oceans are the Pacific, Indian, and Atlantic; the Arctic is the shallowest. The average depth of the oceans is about 10,165 feet (Figure 9).

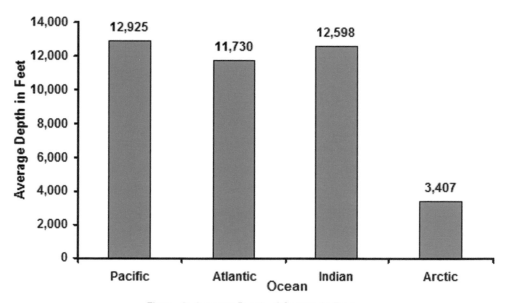

Figure 9. Average Depth of Oceans in Feet

Four billion years ago, Earth could have been described as a very large, hot rock without a trace of water on its surface. Earth scientists believe that during the first few hundred million years after the solar system formed, gases came from within Earth while it differentiated into a core, mantle, and crust. Volcanic activity releases gases from molten rock in the mantle of the planet. Volatile gases that were trapped deep inside the planet were released when the rocks that contained them melted and erupted from volcanoes. Carbon dioxide, methane, sulfur, and other volatiles stayed in their gaseous states to make up the early atmosphere. Erupted water vapor largely condensed to form the early ocean.

Many people use the terms "ocean" and "sea" interchangeably when speaking about the ocean. But geographically, seas are classified as smaller than oceans and are usually located where land and ocean meet. Typically, seas are partially enclosed by land. Also, several "seas," such as the Caspian Sea, are really lakes.

**Seawater** is water from a sea or an ocean. It contains about 3.5 percent salt on average (35 grams per liter), plus smaller amounts of other substances. The salinity of water in major seas varies from about 0.7 percent in the Baltic Sea to 4.0 percent in the Red Sea. The physical properties of seawater differ from those of freshwater in some important respects. Seawater freezes at a temperature that is slightly colder than fresh water (0.0° Celsius). It freezes at a lower temperature (about −1.9°C) because of the salt in it. The density of seawater is much more dependent on temperature, generally decreasing as the temperature increases. However, the density of freshwater is not monotonic function of temperature; water density reaches a maximum at temperature of 3.98°C under normal atmospheric pressure.

**Tides** are the rise and fall of sea levels caused by the combined effects of the rotation of the planet and the gravitational forces exerted by the moon and the sun. Low tides occur at 90-degree angles from these points. High tides occur at the points on Earth's surface that align with the moon. Tides cause changes in the depths of the marine and estuarine waters. Tides also produce oscillating currents known as tidal streams (Figures 10 and 11).

Figure 10. Morning Low Tide: Cala Trebaluger Beach on Menorca, Balearic Islands, Spain

Figure 11. Mid Afternoon High Tide: Cala Trebaluger Beach on Menorca, Balearic Islands, Spain

In other words, the changing tide produced at a given location on Earth is the result of the changing positions of the moon and sun relative to Earth coupled with the effects of Earth's rotation and the local bathymetry (underwater depth of the ocean). The intertidal zone, a strip of seashore that is submerged at high tide and exposed at low tide, is an important ecological product of ocean tides, because it is a region in which two very different types of environment meet, and in which a complex web of interactions takes place.

## ■ 2.2. GLACIERS

A glacier is a huge mass of ice and snow floating in the sea next to land. The effect of gravity on the great weight of the ice makes the glacier move very slowly downhill, its path often ending at the ocean or a lake. A glacier acts like an immense river of ice.

Glaciers form in places where the snow does not melt. A glacier forms as layers upon layers of snow are compacted: as new snow falls, older layers are compressed into dense ice. Most glaciers are found in regions of high snowfall in winter and cool temperatures in summer. This ensures that the snow that accumulates in the winter isn't lost by melt or evaporation during the summer. Such climatic conditions usually prevail in polar and high alpine regions. There are two main types of glaciers: valley glaciers and continental glaciers.

## ■ 2.3. CLOUDS

A cloud is a visible mass of liquid droplets or frozen crystals that float in Earth's troposphere (the lowest part of the atmosphere), moving with the wind. From space, clouds are visible as a white veil surrounding the planet (Figure 12).

Clouds form when water vapor[2] condenses[3] onto microscopic dust particles floating in the air. When cold air meets warm air, the cold air squeezes under the warm air and forces it up, where it expands, cools, and condenses to form clouds. This occurs because cool air can hold less water vapor than warm air, and excess water condenses into either liquid or ice.

---

2  Water vapor is water that has evaporated from the ocean, lakes, and rivers.
3  To condense is to turn into liquid water.

Figure 12.
White Clouds

## ■ 2.4. RAIN AND SNOW

Rain—and other forms of precipitation, such as snow—occur when warm, moist air cools and condensation occurs.

Rain and snow are phenomena that start in clouds: when more water condenses onto other water droplets, the droplets grow. When they get too heavy to stay suspended in the cloud, even with the updrafts within the cloud, they fall to the ground as rain. If the air in the cloud is below the freezing point (0°C), ice crystals form; if the air all the way down to the ground is also freezing or below, we get snow.

## ■ 2.5. GROUNDWATER

In places where there is not enough water above ground, farmers and local water agencies turn to the groundwater found in aquifers to meet their growing demands. These aquifers have become one of the most important natural resources in the world today.

There are many physical obstacles to groundwater. To flow, it has to go through pores in the rocks and soil. Because of this, groundwater flows slower than surface water. The actual rate depends on the transmissivity[4] and storage capacity[5] of the aquifer. Natural outflows of groundwater take place through springs and riverbeds when the groundwater pressure is higher than atmospheric pressure at the ground surface.

## ■ 2.5.1. ORIGIN OF GROUNDWATER

**Groundwater** comprises water that flows under the land surface that supplies water to wells and springs. It is derived largely from precipitation, such as rainfall or snow melt. Precipitation that does not form part of surface runoff or remain on the land surface percolates into the ground. Once there, it can follow any of three distinct paths:

> ❱ Remain in the unsaturated (vadose) zone, subject to capillary action.

> ❱ Return to the atmosphere by evaporation and transpiration.

4   A measure of how much water can be transmitted horizontally, such as to a pumping well.
5   The rate of discharge of water from the well divided by the drawdown of the water level within the well.

> Flow downward until it reaches the water table, joining the groundwater proper.

Depending on its flow path and fate, groundwater flow is classified as:

> **Shallow groundwater.** Typically found near streams, lakeshores, and wetlands. Shallow groundwater is easily exploited and readily recharged (that is, replenished) within weeks, months, or years.

> **Deeper groundwater.** Indirectly connected to the surface environment. More expensive to access and may be recharged very slowly (over decades, centuries, or thousands of years).

## ■ 2.5.2. AGE OF GROUNDWATER

Groundwater in many parts of the world can be thousands to tens of thousands of years old. Generally, groundwater does not recycle as easily as surface water. Rates of groundwater turnover vary from days to years and from centuries to millennia, depending on aquifer location, type, depth, properties (such as temperature, turbidity, and chemical position), and hydraulic conductivity[6] of a soil. The average time for the renewal of groundwater is fourteen hundred years (USSR 1978). Deep groundwater resources show a very long renewal time, while shallow groundwater resources show a very short renewal time. Renewal rates of deep groundwater are about one-fifteenth of those of shallow groundwater (Jones 1997).

## ■ 2.5.3. AQUIFERS

Depending on their position relative to the ground surface and other permeable or impermeable layers (aquitards), aquifers are classified into unconfined and confined aquifers.

**Unconfined aquifers** are where the water table is exposed to the atmosphere through the open pore spaces of the overlying soil or rock. The water table is the upper groundwater surface in an unconfined aquifer; the surface is where the water pressure head is equal to the atmospheric pressure (where gauge pressure is zero). The depth to the water table varies according to

---

6  The soil's ability to transmit water when submitted to a hydraulic gradient.

several factors, such as the topography, geology, seasonal and tidal effects, and the quantities of water being pumped from the aquifer. These aquifers are typically shallow and normally are comprised of surface sands and sandstones. Unconfined aquifers are recharged by direct infiltration of rainwater or stream water from the ground surface. One example of an unconfined aquifer system is a coastal dune system.

**Confined aquifers** have an aquiclude (impermeable layer) on top of the zone of saturation and below it. Water generally enters these aquifers where the permeable rock is present at the surface but is between two types of rock that are not permeable.

When a well is drilled into a confined aquifer, the water that is under pressure in it rises in the well casing and may reach the surface. Wells with water flowing to the surface are often called free-flowing artesian wells. This type of well seems to defy gravity because the pressure that builds up between layers of rock gets relieved when the water finds a path to the open air.

Confined aquifers may be recharged by rain or stream water infiltrating the rock at some considerable distance away from the confined aquifer.

## 2.6. SPRINGS

**A spring** is where groundwater flows naturally onto the land surface or into a body of surface water. The water source of most springs is rainfall that seeps into ground uphill from the spring outlet. A spring recharge basin, or a spring shed, refers to the area within the groundwater basin that contributes to the discharge of a spring.

**A** spring occurs when groundwater appears at the land surface; its behavior depends on the topography, underlying geologic strata, and water levels in the aquifer systems. There are many types of springs:

> Perennial springs flow continuously throughout the year. They derive from extensive permeable aquifers and discharge all the time.

> Intermittent springs flow only during certain times of the year when rainfall or snowmelt is sufficient to recharge the soil and groundwater.

> ❱ Seepage springs are formed when groundwater slowly seeps out of the ground.

> ❱ Artesian springs are those in which water is under pressure and generally flows through fissures.

> ❱ Thermal springs are where groundwater discharge temperatures exceed the regional average air temperature.

## ■ 2.7. LAKES

Lakes are bodies of water surrounded by land (Figure 13). They are larger than ponds and generally are filled with freshwater. Though lakes are a small percentage of Earth's total water, they are important for our economy and transportation.

The largest lakes formed many thousands of years ago when glaciers carved out deep valleys and moved large amounts of soil. When the glaciers melted, they left behind a series of large holes that filled with the melt water. Tectonic movement, volcanic activity, and river erosion formed other lakes. We have also made lakes artificially.

In mountain areas, snow from surrounding mountains drains into a kind of bowl where water pools. When these lakes overflow, water travels down the mountainsides as rivers or streams.

No one is certain how many lakes there are in the world.

The world's biggest lakes in order of size (Figure 14) are:

> ❱ The Caspian Sea, in Eurasia

> ❱ Lake Superior in North America

> ❱ Lake Victoria in Africa

> ❱ The Aral Sea in Asia

> ❱ Lake Huron in North America

Figure 13.
Moraine
Lake, Banff
National
Park, Alberta,
Canada

33

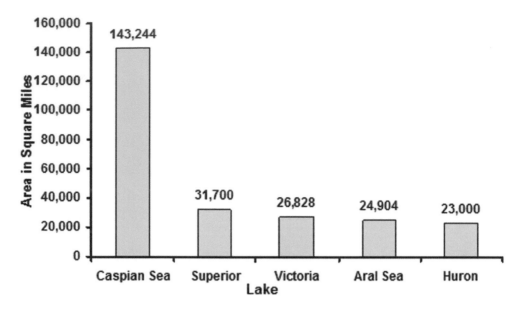

Figure 14. The Areas of the World's Biggest Lakes

### ■ 2.8. RIVERS

Rivers contain freshwater. While rivers cover a small fraction of Earth's total surface, they are extremely important for transportation and the economy (especially agriculture).

Rivers are long, flowing bodies of water. Rivers usually flow toward an ocean, a lake, a sea, or another river. In a few cases, a river simply flows into the ground or dries up completely at the end of its course and does not reach another body of water. A small river may be called a stream.

Rivers also have a significant value in Earth's water cycle as surface water. Water generally collects in a river from precipitation through a drainage basin from surface runoff and other sources, such as groundwater recharge, springs, and the release of stored water in natural ice and snow packs (e.g., from glaciers).

The most famous and largest rivers by length (Figure 15) are:

❭   The Nile in Africa

❭   The Amazon in South America

> The Mississippi/Missouri in North America

> The Ob in Asia

> The Yangtze/Kiang in Asia

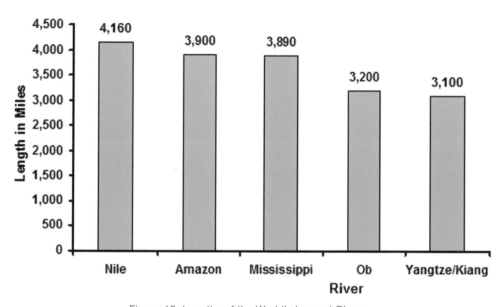

Figure 15. Lengths of the World's Largest Rivers

The ages of surface water and groundwater are very different. Surface waters are typically fresh and distinctly new. Globally, surface water recycles every nine to sixteen days, with an average of eleven days (L'vovich 1979).

## ■ 2.9. OTHER BODIES OF WATER

While oceans, lakes, and rivers are the major types of bodies of water, there are other types.

A **gulf** is a body of water that can be found at the mouth of a sea. The Gulf of Mexico, a part of the Atlantic Ocean by the southern United States and Mexico, is an example.

A **bay** is a part of an ocean or lake extending into the land. It is similar to a gulf but usually smaller.

A **cove** is a small indentation or recess in the shoreline of a sea, lake, or river.

A **sound** is a narrow body of water separating an island from the mainland.

A **strait** is a narrow channel of the sea joining two larger bodies of water. The Bering Strait between Alaska and Russia is a good example.

A **waterfall** is a cascade or a steep fall of the water of a river or stream.

A **dam** is any barrier that holds back water in a lake, river, stream, or other water body. Dams are primarily used to save, manage, and/or prevent the flow of excess water into specific regions. Moreover, some dams are used to generate hydropower.

Figure 16 shows Alwen Reservoir, which is a 5km long reservoir near Pentre-Llyn-Cymmer in the county borough of Conwy, North Wales. It is held back by the 27 metre high Alwen Dam. It was constructed between 1909 and 1921 to supply water to the town of Birkenhead, near Liverpool in England. Today the reservoir supplies water to homes across north-east Wales producing about 5 million gallons of water a day.

## 3. CONCLUSION

Water is the only substance on Earth that exists simultaneously in nature as a solid, a liquid, and a gas. Liquid is the most common of water's states and is the form that is generally denoted by the world "water." Water is found on Earth as oceans, lakes, groundwater, rivers, and in other forms. Solid water is water that is frozen and found largely as glaciers at both north and south poles. In many places in the world, when the air temperature is less than zero degrees Celsius, water takes the form of snow. Water is a gas when it is in the atmosphere as vapor or steam; a great amount of it in the air may take the form of clouds.

Figure 16. Alwen Reservoir Dam, North Wales, United Kingdom

# CHEMICAL AND PHYSICAL PROPERTIES OF WATER

## 1. INTRODUCTION

Water is the most abundant molecule on Earth's surface. We've seen that water is unique as the only substance that occurs in three states simultaneously in nature: solid (ice), liquid, or gas (vapor). Let's look at the chemical and physical properties of water.

## 2. CHEMICAL AND PHYSICAL PROPERTIES OF WATER

Water's chemical description is $H_2O$. The hydrogen atoms are attached to one side of the oxygen atom, resulting in a positive charge on the side where the hydrogen atoms are and a negative charge on the other side, where the oxygen atom is. This opposition of charges is called polarity (Figure 17). Water molecules exist in liquid form over an important range of temperature: from 0° to 100° Celsius, permitting life to survive in climate and weather changes. This range allows water molecules to exist as a liquid in most places on the planet. Earth's water is constantly interacting, changing, and moving.

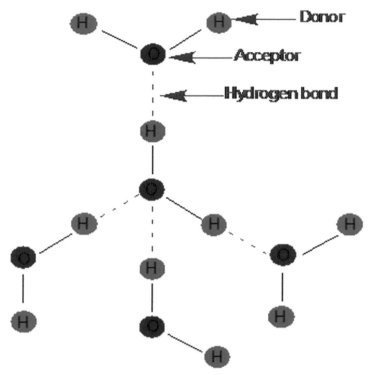

Figure 17. Hydrogen Bonding of Water Molecules

## ■ 2.1 WATER, ICE, AND VAPOR

As we have seen, water vapor is the gas phase of water. It can be produced from the evaporation or boiling of liquid water or from the sublimation[7] of ice.

## ■ 2.1.1 THE SPECIFIC HEAT CAPACITY OF WATER

Water has a high specific heat index due to hydrogen bonding, which increases intermolecular forces between molecules. This means that water absorbs a lot of heat before it gets hot. Therefore, it is greatly responsible for regulating extremes in our environment. This is why water is valuable to industries and in your car's radiator as a coolant. The high specific heat index of water also serves to buffer the internal temperatures of organisms and helps to regulate the rate at which air changes temperature. That is why the temperature change between seasons is gradual rather than sudden, especially near the oceans.

---

7  Ice turns directly into water vapor without first transitioning into a liquid.

## ■ 2.1.2 DENSITY OF WATER AND ICE

The density of water is the weight of the water per its unit volume, which depends on the temperature of the water. With increase and decrease of temperature, the volume increases or decreases, changing the water's density. The maximum density of water occurs around 4°C (Figure 18).

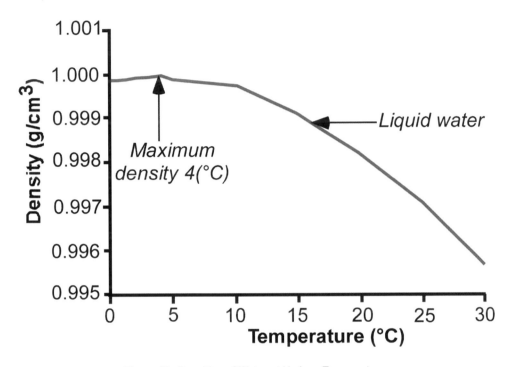

Figure 18. Densities of Water at Various Temperatures

Ice floats in water because solid water (ice) is less dense than liquid water. You may have noticed that when you add ice to a glass of water, the ice floats (Figure 19).

The density of ice is 0.9167 g/cm³, while the density of water is 0.9998 g/cm³. Ice floats because it is about 9 percent less dense than liquid water, which also means it takes up about 9 percent more space than water. The heavier water displaces the lighter ice, so ice floats to the top. Water in either solid or liquid states forms hydrogen bonds because of the polar nature of the $H_2O$ molecule (Figure 17). In the frozen state, the molecules are held together by the hydrogen bonds in a crystalline structure.[8] In it, there is much more space between each molecule than in the liquid state, which

---

8  The configuration in which atoms are arranged in a material.

makes ice less dense than water. When ice melts, many, but not all, of the bonds are broken, but the regularity of the crystalline arrangement is broken up and the space between the molecules decreases. Liquid water has loosely bound molecules.

Figure 19. Iceberg in Paradise Bay, Antarctica

Ice turns into vapor after the intermediate liquid state known as melting (solid to liquid phase transition). In addition, ice can turn directly into vapor without first transitioning into a liquid, it is referred to as sublimation. Figure 20 shows melting icebergs in Jokulsarlon Lagoon, Iceland.

Given air and water at the same temperature, an ice cube melts faster in water. You can test this by melting an ice cube in a cup of water. The ice is exposed to both air and water, but you will see that the part of the ice cube in the water melts faster than the part in the air.

Ice melts faster in water than in air (Figures 20 and 21) because:

> The molecules in liquid water are more tightly packed than the molecules in air, allowing more contact with the ice and a greater rate of heat transfer.

> Water has a higher heat capacity than air. This means that heat can travel through the water to reach the ice more easily than it can through the air.

## ■ 2.1.3 DENSITY OF SALT WATER AND ICE

**S**eawater has a higher density than freshwater. The density of ice is 920 kg/$m^3$, and that of seawater is 1,030 kg/$m^3$. Seawater (which is salt water) has a higher mineral content than freshwater; it contains many dissolved substances that add mass to it. Therefore, it has a greater mass per unit volume—that is, a density higher than that of freshwater.

## ■ 2.1.4 MISCIBILITY AND CONDENSATION

**Miscibility** is the property of liquids (and particularly of water) to mix in all proportions, forming a homogeneous solution.

**Condensation** is the process by which water vapor in the air is changed into liquid water. It is also the process that creates clouds, rain, and snow.

Figure 20. Melting Icebergs in Jokulsarlon Lagoon, Iceland

Figure 21.
Icicle on
Icy Air
Conditioner

44

## ■ 2.1.5 VAPOR PRESSURE

Vapor pressure is the pressure of a vapor in thermodynamic equilibrium with its condensed phases in a closed container. It is technically the pressure of water vapor above a surface of water; it is always present in the air, and it fluctuates. When air reaches the saturation vapor pressure, the water vapor in it condenses. The temperature at which air containing a certain amount of water vapor becomes saturated is called the dew point.

The vapor pressure of liquid water varies with its temperature. As the temperature of liquid (or solid) water increases, its vapor pressure also increases (Figure 22).

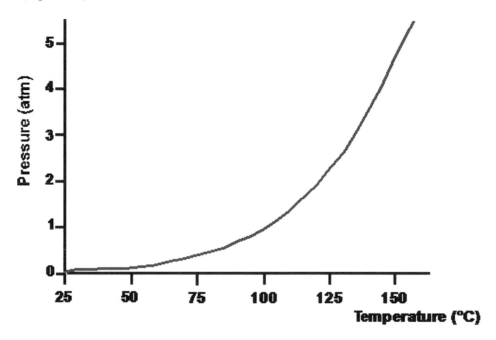

Figure 22. Vapor Pressure of Liquid Water

## ■ 2.1.6 COMPRESSIBILITY OF WATER

The compressibility of water is low. It is a function of pressure and temperature. At 0°C, at the limit of zero pressure, the compressibility is $5.1 \times 10^{-10}$ $Pa^{-1}$ ((Pa) pascals = 1 atmosphere; 105 pascals = 1 bar) (Fine and Millero 1973). At the zero-pressure limit, the compressibility reaches a minimum of $4.4 \times 10^{-10}$ $Pa^{-1}$ around 45°C before increasing again with increasing temperature. An increase of pressure by 1 atmosphere (= 1013mbar = 14.7 psi) causes a decrease of the water volume by $5.3 \times 10^{-5}$ of the original volume.

### ■ 2.1.7. TRIPLE POINT

The triple point is the intersection on a phase diagram where three phases coexist in equilibrium. The gas–liquid–solid triple point of water corresponds to the minimum pressure at which liquid water can exist. The triple point of water, the single combination of temperature and pressure at which liquid water, solid ice, and water vapor can coexist in a stable equilibrium, occurs at exactly 273.16 K (0.01°C) and a partial vapor pressure of 611.73 pascals. At that point, it is possible to change all of the substance to vapor, water, or ice by making arbitrarily small changes in pressure and temperature (Figure 23).

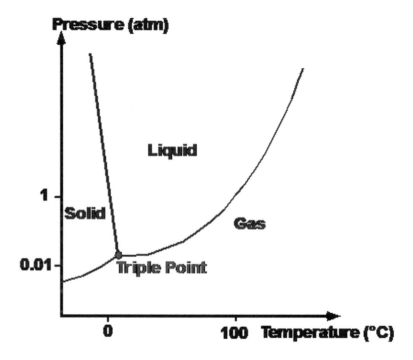

Figure 23. Phase Diagram

At pressures below the triple point, solid ice when heated at constant pressure sublimates directly into water vapor. However, at pressure above the triple point, solid ice when heated at constant pressure first melts to form liquid water and then evaporates or boils to form vapor at a higher temperature. The triple point of water is sometimes used in the calibration of measurement equipment.

### ■ 2.2 ELECTRICAL PROPERTIES OF WATER

Water also has electrical properties: electrical conductivity and electrolysis.

## ■ 2.2.1 ELECTRICAL CONDUCTIVITY

Conductivity is a measure of the ability of water to pass an electrical current. It estimates the total amount of solids dissolved in water as expressed by "total dissolved solids" (TDS). These dissolved solids can include chloride, nitrate, sulfate, and phosphate anions (ions that carry a negative charge), or sodium, magnesium, calcium, iron, and aluminum cations (ions that carry a positive charge). TDS is measured in parts per million (ppm) or in milligrams per liter. The amount of conductivity is directly proportional to the amount of salts dissolved[9] in the water and thus, water shows significant conductivity when dissolved salts are present. However, pure water is a poor conductor of electricity. In addition, conductivity varies with temperature: the higher the temperature, the higher the electrical conductivity. For this reason, conductivity is reported as conductivity at 25 degrees Celsius (25°C).

## ■ 2.2.2 ELECTROLYSIS

**Electrolysis of water** is the decomposition of water ($H_2O$) into oxygen ($O_2$) and hydrogen gas ($H_2$) when electric current is passed through it. The electrolysis of one mole[10] of water produces a mole of hydrogen gas and a half-mole of oxygen gas in their normal diatomic forms.[11]

## ■ 2.3 DIELECTRIC CONSTANT

The dielectric constant, or permittivity - ε - is a dimensionless constant that indicates how easily a material can be polarized by imposition of an electric field on an insulating material. The dielectric constant is equal to the ratio of the capacitance[12] of a capacitor[13] filled with the given material to the capacitance of an identical capacitor in a vacuum[14] without the dielectric[15] material. The dielectric constant can be expressed as: $\varepsilon = \varepsilon_s / \varepsilon_0$

---

9  Composed of related numbers of cations (positively charged ions) and anions (negative ions).

10  The amount of pure substance containing the same number of chemical units as there are atoms in exactly 12 grams of carbon-12 (i.e., 6.023 X 1023).

11  Molecules made of two atoms chemically bonded together.

12  Capacitance is the ability of a body to store an electrical charge.

13  A capacitor is a passive, two-terminal electrical component used to store energy electrostatically in an electric field.

14  A vacuum is space that is devoid of matter.

15  An electrical insulator that can be polarized by an applied electric field.

Where:

$\varepsilon$ = the dielectric constant

$\varepsilon_s$ = the static permittivity of the material

$\varepsilon_0$ = vacuum permittivity

The value of the dielectric constant at room temperature (25°C) is 78.2 for water.

## ■ 2.4 POLARITY AND HYDROGEN BONDING

The water molecule, as a whole, has ten protons (each of which carries a positive charge) and ten electrons (each of which carries a negative charge), so it is neutrally charged. However, the $H_2O$ molecule has polarity. The oxygen atom and hydrogen atoms share electrons in covalent bonds,[16] but the sharing is not equal. The central oxygen atom is more electronegative than the two hydrogen atoms, and so the electrons shared in the two bonds spend more time around the oxygen atom than around the hydrogen atoms. This structure is shown in Figure 17.

In the covalent bond between oxygen and hydrogen, the oxygen atom attracts electrons a bit more strongly than the hydrogen atoms do. The force of attraction, shown here as a dotted line, is called a hydrogen bond. Each water molecule is hydrogen-bonded with up to four others.

The hydrogen bonds that form between water molecules give water molecules two additional characteristics:

   ❯   The attraction created by hydrogen bonding keeps water in
        a liquid phase over a wider range of temperatures than for
        any other molecule its size.

   ❯   The energy required to break multiple hydrogen bonds causes
        water to have a high heat of vaporization. Therefore, a large amount
        of energy is required to convert liquid water to water vapor.

---

16   A chemical bond formed by the sharing of one or more electrons, especially pairs of electrons, between atoms.

### ■ 2.4.1 COHESION AND ADHESION

The hydrogen bond between water molecules is the reason behind two of water's unique properties: **cohesion** and **adhesion**.

**Cohesion** refers to the fact that water sticks to itself very easily because water molecules are attracted to each other.

**Adhesion** means that water also sticks very well to other things because water molecules are also attracted to molecules of other substances.

### ■ 2.4.2 SURFACE TENSION

Water's high surface tension is a result of hydrogen bonding and cohesion between the water molecules. The molecules cohere to each other strongly but adhere to the other medium weakly. All forms of water (rain, water in a glass, etc.) are examples of cohesion. For example, a drop of water doesn't fall from the rim of a faucet immediately but will stretch itself as thinly as possible first. As it falls, it then forms a spherical shape, and no other.

### ■ 2.4.3 CAPILLARY ACTION

Capillary action is important for moving water around. It is defined as the movement of water within the spaces of a porous material due to the forces of adhesion, cohesion, and surface tension. For example, capillary action happens in plants when they "suck up" water. The water adheres to the insides of the plant's tubes, but surface tension attempts to flatten it out. This makes the water rise and cohere to itself again, a process that continues until enough water builds up to make gravity pull it back down.

### ■ 2.4.4 WATER AS A SOLVENT

Water is often known as the **universal solvent** because more substances dissolve in water than in any other chemical. Although the water molecule carries no net electric charge, its eight electrons are not distributed uniformly; as we've noted, there is slightly more negative charge (purple) at the oxygen end of the molecule and a compensating positive charge (green) at the hydrogen end (Figure 17). The resulting polarity is largely responsible for water's unique properties, particularly its ability to dissociate ionic compounds into their positive and negative ions. The positive part of an ionic compound is attracted to the oxygen side of water, while the negative portion of the compound is attracted to the hydrogen side. The substances that dissolve in water are known as **hydrophilic**. Like water, salt and sugar are both polar, so they dissolve very well in it. Substances that do not dissolve in water are known as **hydrophobic.** (You've heard the saying, "oil and water don't mix.")

## ■ 2.5 WATER IN ACID-BASE REACTIONS

Water is formed by an $H^+$ cation, an acid,[17] combined with an $OH^-$ anion, a base.[18] An **acid-base reaction** is a chemical reaction that occurs between an acid and a base.

> ❯ In the Bronsted definition, it is the transfer of $H^+$ from acid to base.

> ❯ In the Lewis definition, it is the transfer of an electron pair from base to acid.

The pH scale (potential hydrogen) measures how acidic or basic a substance is (pH refers to the amount of hydrogen mixed in with the water). The pH scale ranges from 0 to 14. A pH of 7 is neutral, which applies to pure water and distilled water. Most water, however, is not exactly pure. Acidic water is water with a pH of less than 7. An example of an acid is vinegar, with a pH of 3. A pH greater than 7 is basic (alkaline). An example of an alkaline is milk of magnesia solution, with a pH of 10 (Figure 24).

**N**ormal rain has a pH of about 5.6. **Acid rain** is not pure acid falling from the sky; it is normal rain that reacts with elements and gases in the atmosphere, such as carbon dioxide, forming mildly acidic carbonic acid before it becomes rain. Acid rain has a pH of about 5.0 or less and can even be in the 4 range in the northeastern United States, where there are a lot of industries and cars and where limestone does not naturally occur in the soil. (Limestone's soils have a pH of greater than 7 and tend to balance out some of the acid rain's acidity.) Acid rain can harm the environment. **Surface waters** and their fragile ecosystems are perhaps the most famous victims of acid rain. The precipitations infiltrate lakes, rivers, and streams through

17  An acid is a substance that donates hydrogen ions.
18  A base is a substance that accepts hydrogen ions.

soils that can resist changes in acidity and alkalinity; this can cause a water body's acidity to increase by many times its normal level. Plankton and invertebrates are very sensitive to changes in acidity and die first, damaging the base of the food chain. Some fish and animals, such as frogs, have a hard time adapting to and reproducing in an acidic environment. Many plants, such as evergreen trees, are damaged by acid rain and acid fog, so acid rain also damages forests, such as the Black Forest of Germany.

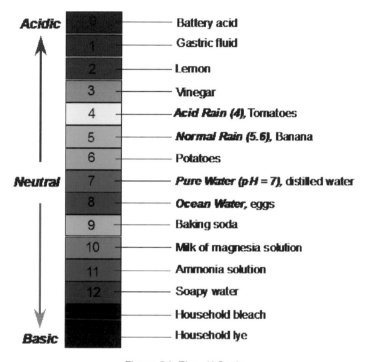

| | |
|---|---|
| **Acidic** | 0 — Battery acid |
| | 1 — Gastric fluid |
| | 2 — Lemon |
| | 3 — Vinegar |
| | 4 — *Acid Rain (4)*, Tomatoes |
| | 5 — *Normal Rain (5.6)*, Banana |
| | 6 — Potatoes |
| **Neutral** | 7 — *Pure Water (pH = 7)*, distilled water |
| | 8 — *Ocean Water*, eggs |
| | 9 — Baking soda |
| | 10 — Milk of magnesia solution |
| | 11 — Ammonia solution |
| | 12 — Soapy water |
| | — Household bleach |
| **Basic** | — Household lye |

Figure 24. The pH Scale

The body's pH balance is very important. Proper health starts with the correct acid-alkaline balance. There are special filters and machines that you can buy to help streamline your production of alkaline acid water.

Water with some acid and alkaline properties has many beneficial uses for us:

❭ **Acidic ionized water** can be used externally for cleaning and disinfecting the skin and household surfaces.

    ❭ Daily skin care: It can improve the complexion of your skin if you wash with it. It works as a natural astringent and removes dirt and oil.

❯ Skin problems: It helps to relieve dry and itchy skin and contributes to clearing acne, psoriasis, eczema, and athlete's foot.

❯ Hair care: It is great for your scalp and addresses dandruff.

❯ Antiseptic: It is ideal for hand washing and expedites the healing of cuts, minor wounds, insect bites, rashes, and fungal infections.

❯ Sterilization: Strong acidic water (with a pH under 3.0) kills pathogens on contact and can be used to sterilize cutting boards, kitchen utensils, and other household surfaces.

❯ Food: Wash your fruits, vegetables, and meats to kill bacteria.

❯ Plant care: Watering plants and vegetables with acidic water promotes growth and reduces fungus. Cut flowers last longer in acidic water.

❯ **Alkaline water** is important for optimum health.

❯ Improves blood: It assists oxygenation in the bloodstream and allows optimum cleansing and detoxification.

❯ Increases energy: You can feel the difference, usually within a few days.

❯ Increases intracellular hydration of cells and skin and replenishes minerals.

❯ Its detoxification effect may slow the aging process and help to prevent diseases caused by extended acidic pH levels.

> ❱ Alkaline water has powerful antioxidants. Electron-charged water helps control free radical activity.

> ❱ Improves digestion and metabolism and lowers blood sugar levels

## ■ 2.6. WATER IN OXIDATION-REDUCTION REACTIONS

Oxidation-reduction reactions (also known as "redox") are a type of reaction in aqueous solutions that involves a transfer of electrons between two chemical species. Oxidation and reduction have to occur together—an oxidation reaction must have a corresponding reduction reaction.

**Oxidation** involves the loss of electrons or hydrogen, a gain of oxygen, or an increase in oxidation state. **Reduction** involves the gain of electrons or hydrogen, a loss of oxygen, or a decrease in oxidation state.

Oxidation-reduction reactions are used in industry to reduce ores to obtain metals, to produce cleaning products, to convert ammonia into nitric acid for fertilizers, and to coat compact discs. Redox reactions are vital for biochemical reactions, and they take place in electrochemical cells and in respiration and photosynthesis—basic life functions.

## ■ 2.7 TRANSPARENCY

Transparency is a measure of how clear water is. Water molecules don't absorb or reflect most visible light; instead, they allow it to pass through relatively unimpeded. However, water absorbs wavelengths like infrared and reflects invisible UV. Transparency is very important, because aquatic plants need sunlight for photosynthesis. The clearer the water, the deeper sunlight will penetrate (Figure 25).

## ■ 2.8 HEAVY WATER AND ISOTOPOLOGUES

"**Heavy water**," formally called deuterium oxide ($2 H_2O$ or $D_2O$), is a form of water that contains a larger than normal amount of the hydrogen isotope deuterium (also known as "heavy hydrogen") rather than the common hydrogen-1 isotope that makes up most of the hydrogen in normal water (IUPAC 1997).

Figure 25. Big Rock in Water Showing Water Transparency, Mediterranean Beach in Mallorca, Spain

**Isotopologues** are chemical species that differ only in the isotopic composition of their molecules or ions. When deuterium oxide mixes with water, three isotopic forms occur: HOH, HOD, and DOD. These isotopologues have at least one atom with a different number of neutrons. Water's hydrogen-related isotopologues also include "light water" (HOH or $H_2O$) and "semiheavy water," with the deuterium isotope in equal proportion to protium (HDO or 1H2HO).

## 3. CONCLUSION

Water has a number of unique chemical and physical properties that life requires to start and to continue. If one of these properties is disturbed, it could threaten life. Indeed, water and life are inseparable. To understand the relationship between water and life better, let's look at how water circulates.

# CIRCULATION OF WATER

## 1. INTRODUCTION

The circulation of water above and below the surface of Earth is also known as the **water cycle** or **hydrologic cycle**. The circulation of the waters of the hydrosphere results in the weathering of the landmasses (Hanor 2002). Since the water cycle is truly a "cycle," there is no beginning or end. Individual water molecules can come and go, in and out of the atmosphere.

This chapter focuses on the circulation of Earth's water and on the various processes associated with running water.

## 2. HEAT FROM THE SUN

The sun is a huge sphere of glowing gases that produces energy and light, making life on Earth possible. The sun, with its heat energy, has a huge effect on water circulation. The sun warms our planet, heating the surface, the oceans, and the atmosphere. This energy to the atmosphere is one of the primary drivers of weather (Figure 26).

The sun warms the oceans around the tropics, and its absence cools the water around the poles. Because of this, the oceans have warm and cold surface currents that act like a global heating and air-conditioning system, drastically affecting the weather and climate around the world.

The sun also drives the **water cycle**. While the sun warms the surface of tropical waters to 80°F (26.67°C), at depths of five thousand feet, temperatures are near freezing. This temperature differential is key for the open-cycle ocean thermal energy conversion (OTEC) process, in which warm surface water is pumped to a vacuum chamber to produce steam that drives a turbine. At the same time, a heat exchanger uses cold water pumped from the depths to condense spent steam into drinkable water (Valenti 1996).

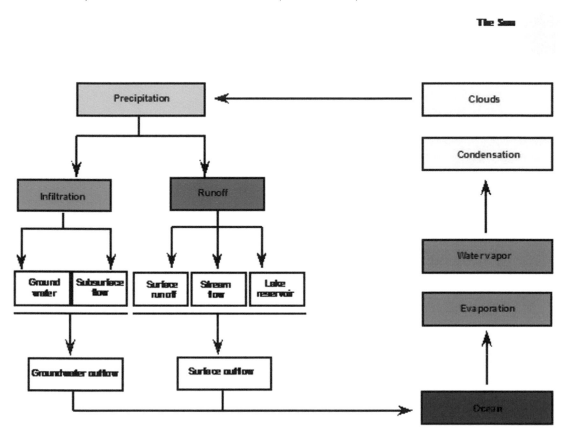

Figure 26. Circulation of Water in the "Water Cycle"

# 3. EVAPORATION FROM SEA AND LAND

**Evaporation** is the transformation of water from the liquid to the gas phase, thus transferring water from land and water masses to the atmosphere (Figure 26). Water evaporates due to heat from the sun. Evaporation controls the loss of freshwater. Evaporation from the sea surface is important in the movement of heat in the climate system; 86 percent of global evaporation occurs over the ocean (Baumgartner and Reichel 1975). Water evaporates from the surface of the ocean, mostly in warm, cloud-free, subtropical seas, and cools the environment. The large amount of heat absorbed by the ocean partially buffers the greenhouse effect from increasing carbon dioxide and other gases.

**Transpiration** is the biological process of water movement that occurs mostly in the day. Water inside plants is transferred from them to the atmosphere as water vapor through aerial parts—especially leaves, but also stems and flowers.

**Evapotranspiration** is the sum of evaporation and plant transpiration from Earth's land surface to the atmosphere. Total annual evapotranspiration amounts to approximately 505,000 km$^3$ (121,000 cubic miles) of water, 434,000 km$^3$ (104,000 cubic miles) of which evaporates from the oceans (Swarthout and Hogan 2010).

# 4. CONDENSATION

Condensation is the process by which water vapor changes its physical state from a vapor to a liquid (most commonly), when it touches a cooler surface. This is the opposite of evaporation (Figures 19, 20, 26, and 27).

Water vapor carried by the atmosphere condenses as clouds and falls as precipitation, the primary way water returns to the surface. The most active particles that form clouds are sea salts, atmospheric ions caused by lightning, and combustion products containing sulfurous and nitrous acids. Condensing water vapor releases latent heat, and this drives much of the atmospheric circulation in the tropics. This latent heat release is an important part of Earth's heat balance, and it couples the planet's energy and water cycles.

# 5. PRECIPITATION AS RAIN AND SNOW

Precipitation is a form of water (such as rain, snow, or sleet) that condenses from a gaseous state in the atmosphere, becomes too heavy to remain suspended, and then falls to Earth's surface (Figure 26). Most global precipitation occurs over the ocean: 78 percent (Baumgartner and Reichel 1975). Precipitation also affects the height of the ocean surface indirectly via salinity and density. Depending on the air temperature, the water droplets can return to Earth in either a liquid form (as rain) or in a solid form (as snow, sleet, or hail). Approximately 505,000 km$^3$ (121,000 cubic miles) of water falls as precipitation each year—398,000 km$^3$ (95,000 cubic miles) of it over the oceans (Poehls and Smith 2011).

When the water returns to Earth, it either infiltrates the surface and collects underground in aquifers or becomes runoff that flows into rivers and streams. Precipitation governs most of the gain of freshwater.

Figure 27.
Water Drops
on Glass
Surface

# 6. INFILTRATION

**Infiltration** occurs when water falls and soaks into the ground. Once it infiltrates, we call the water soil moisture or groundwater (Figure 26). Two forces govern infiltration: gravity and capillary. High infiltration rates occur in dry soils, with infiltration slowing as the soil becomes wet. Vegetation also affects infiltration. For instance, infiltration is higher for soils under forest vegetation than for bare soils.

**Subsurface flow** or **hypodermic flow** is the flow of water beneath Earth's surface as part of the water cycle. Partly infiltrated rainfall circulates more or less horizontally in the superior soil layer and appears at the surface through drain channels.

# 7. RUNOFF

Runoff is when water does not soak into the ground but flows into streams and rivers (Figure 26). Runoff includes a variety of ways by which water moves across the land, including:

) **Surface runoff** is the water flow that occurs when the soil is infiltrated to full capacity and excess water from rain, snowmelt, or other sources flows over the land surface. It is a major component of the water cycle and the primary agent in water erosion (Horton 1933, Beven 2004).

) **Streamflow**, also known as channel runoff, is the volume of water that moves through a specific point in a stream during a given period. Water flowing in channels comes from surface runoff from adjacent hill slopes, from groundwater outflow, and from water discharged from pipes. Streamflow is the main mechanism by which water moves from the land to the oceans or to basins of interior drainage.

As it flows, water may infiltrate into the ground, evaporate into the air, become stored in lakes or reservoirs, or be extracted for agricultural or other human uses.

## 8. CONCLUSION

The water cycle figures significantly in the maintenance of life, society, and ecosystems on Earth. The water cycle is the process by which our freshwater is produced. Water transfers from one reservoir to another, such as from river to ocean or from the ocean to the atmosphere, by the physical processes of evaporation, condensation, precipitation, infiltration, runoff, and subsurface flow. Water can change states among liquid, vapor, and ice at various places in the water cycle. These processes can happen over millions of years.

# Part II:

## WORK DONE BY WATER

# WEATHERING OF ROCKS

## 1. INTRODUCTION

Weathering refers to the group of physical, chemical, and biological processes that change the physical and chemical states of rocks and soils at or near Earth's surface. The exact way in which weathering occurs in any particular situation depends primarily on three different factors:

> **Climate.** Weathering is primarily a result of precipitation and temperature forces. High temperatures and greater rainfall increase the rate of chemical weathering.

> **Length of exposure.** Because the effects of climate occur at Earth's surface, the longer a rock is exposed to the agents of weathering, the greater the degree of alteration, dissolution, and physical breakup.

> **Soil.** Rocks covered by soil are subjected to chemical reactions with water much longer than rocks not covered by soil.

The objective of this chapter is to discuss the role of water in weathering processes and the different types of weathering that can affect a rock.

## 2. PHYSICAL WEATHERING

Physical weathering is the disintegration of rock into smaller pieces by mechanical forces concentrated along rock fractures without changing the chemical composition of the rock (Figure 28). For example, gravity may cause a large boulder to break loose from the top of a mountain and fall. When the boulder strikes solid ground, it may break apart into many smaller pieces.

Figure 28. Physical Weathering by Water: The Forest Creek, Gljun, Bovec, Slovenia

There are many types of physical weathering, but some of the most common are:

> **Frost action**, the most common physical weathering process. Frost is widespread throughout the world. Frost action and ice wedging break up rock through the freezing and thawing of water. Because water expands when it freezes, it can break rocks apart from the inside when it seeps into cracks in a rock or soil. A very similar process occurs on roads, causing potholes.

> **Abrasion**, which occurs when water or wind carries particles from rocks, wearing them away. Moving particles in water cause weak and loose material on the rock to dislodge. Broken particles may be carried away into the water or wind, or fall due to gravity. Examples of abrasion are the grinding action of glaciers, as well as gravel, pebbles, and boulders moved along and constantly abraded by fast-flowing streams.

> **Exfoliation**, a form of mechanical weathering in which curved plates of rock are stripped from rock below. This type of weathering is common in warm areas. As natural or human-caused forces remove overlying rocks and soils, underlying rocks may work their way to Earth's surface. As the sun shines on rocks during the day, it causes them to expand. At night, when the temperature is lower, the rock contracts. This continued process causes small rock pieces, often flaky in appearance, to form.

## 3. CHEMICAL WEATHERING

Chemical weathering of rocks or soils on the surface occurs when rocks or soils chemically react with water, solutions, and gases. Chemical reactions result in a change in the composition of a rock.

These chemical processes need water and occur more rapidly at higher temperatures. Warm, damp climates are best. The main agents of chemical weathering are oxygen, rainwater (acid rainfall has accelerated the weathering of human-made objects, including statues, bridges, and buildings), carbon dioxide, and acids produced by decaying plants and animals that lead to the formation of soil.

There are different types of chemical weathering. The most important are:

)   **Chemical oxidation**, the breakdown of rock by oxygen dissolved in water. This reaction often gives iron, the most commonly oxidized mineral element. $Fe^{2+}$ (ferrous iron) $\longrightarrow Fe^{3+}$ (ferric iron) or $2FeO + O_2 \longrightarrow Fe_2O_3$. Oxidation is among the most immediate chemical weathering processes. A change in the color of iron, such as "rusting," often indicates that oxidation has occurred. Other readily oxidized mineral elements include magnesium, sulfur, aluminum, and chromium.

)   **Hydrolysis**, the most common weathering process, in which a compound reacts with water to form one or more new substances. The water molecule splits to form positively charged hydrogen ($H^+$) and negatively charged hydroxide ($OH^-$) ions. Decomposition of minerals in water occurs when hydrogen ions ($H^+$) replace cations in minerals. Pure water is a poor hydrogen donor; however, carbon dioxide ($CO_2$) dissolves in water to produce carbonic acid: $CO_2 + H_2O \longrightarrow H_2CO_3$ (carbonic acid) $\longrightarrow H^+ + HCO_3^-$ (bicarbonate). The most common weathering reaction on Earth is the hydrolysis of feldspars, which produces clay minerals.

)   **Carbonation**, which occurs when carbon dioxide in the air reacts with water to form carbonic acid, as in the reaction described above. Carbonic acid is a weak acid. It partially ionizes to form hydrogen ions through the reaction $H_2CO_3 \Leftrightarrow H^+ + HCO_3^-$. Carbonic acid can attack many kinds of rocks, changing them into other forms. Over millions of years, this very dilute acidic solution has been responsible for the formation of caves in areas of limestone rocks (limestone is mainly composed of calcium carbonate, which reacts with acids to produce carbon dioxide, a salt, and water).

## 4. BIOLOGICAL WEATHERING

Biological weathering refers to the breakdown or degradation of rock by living organisms. For example, a seed falls into a small crevice in a rock and germinates. As the plant grows and sends down roots, it works its way into the rock and expands the crevice. Eventually, the plant's roots may actually tear the rock apart (Figure 29).

Figure 29. Biological Weathering: Giant exposed roots of Sprung trees on Ta Prohm temple at Angkor in Siem Reap province, Cambodia

## 5. CONCLUSION

Water is the key element in the process of weathering, which occurs when water seeps into cracks in rocks and moves through and around rocks and minerals. There are three types of weathering: physical, chemical, and biological. Without weathering, there would just be big rocks everywhere.

# 7

# EROSION OF THE LAND SURFACE

## 1. INTRODUCTION

Erosion is the process of removing material from the Earth's surface. The material produced by erosion is called sediment (made up of sedimentary particles, or grains). Erosion is responsible for the creation of hills and valleys. Erosion also requires a medium, such as water or ice, to move material.

The energy for erosion comes from several sources. Gravity acts to move materials of higher relief vertically to lower elevations, producing equilibrium. The sun drives the hydrologic cycle and is the most common source of heat energy. Earth's rotation drives weather in the form rain or snow and wind. Water erosion is caused by the kinetic energy of rain falling on the soil surface and the mechanical force of runoff. These forces work together to break down and carry away the surficial materials of Earth.

This chapter describes erosion caused by the water flow and seepage that occurs through ground soil and some structures constructed with it.

## 2. DENUDATION BY RAIN AND RIVERS

The wearing away and removal of solid matter from Earth's surface by natural agencies such as rain and rivers is called denudation (erosion). It is the breakdown of rock into smaller pieces through weathering processes. Mountains are rained on and become hills. The pieces of the mountain are eroded to smaller pieces and go down the sides of hills. Denudation, in fact, always happens in a downhill direction.

Factors affecting denudation include:

> Surface topography

> Geology

> Climate (most directly in chemical weathering)

> Tectonic activity

> The biosphere (living organisms and their environments)

> Anthropogenic (human) activity

# 3. WATERFALLS AND GORGES

**A waterfall** is a feature of erosion found in the youth stage of a river; it is a steep drop in its course. A riverbed includes areas with bands of hard (resistant) and soft (less resistant) rock. Since the hard rock takes longer to erode than the soft rock, the river erodes the land at different rates. A waterfall forms when a band of hard, resistant rock (cap rock) lies over softer, less resistant rock affected by the main processes of erosion, including hydraulic action (the force and power of the moving river) and abrasion[19] (Figures 30 and 31).

Three main processes create waterfalls:

> A recent uplift or down-dropping of part of Earth's crust

> Diversion of a river by blockage of a preexisting channel

> Differential erosion of valleys, especially in glaciated areas

---

19 Abrasion is the scraping of the load against the bed and banks.

**A gorge** is a narrow valley with steep, rocky walls located between hills or mountains.

Most gorges are formed through water erosion. Streams carve through hard layers of rock, breaking down or eroding them. Sediment from the worn-away rock is then carried downstream. Erosion may also be caused by "rejuvenation" when a river begins to cut downward into its channel. Over time, this kind of erosion forms the steep walls of a gorge. The flooding of streams or rivers increases the speed and intensity of such erosion, creating deeper and wider gorges.

## 4. SEDIMENTS CARRIED BY RIVERS

Water can move more sediment than air can, because water has both a higher density and viscosity.

The flow of rivers is very important to Earth because rivers are major forces that shape the landscape and transport an incredible amount of sediment. Large pieces of sediment, like sand and gravel, are typically deposited in a fast-moving river, while smaller particles of sediment are carried away.

Figure 30. Maeya Waterfall Erosion: Looking Upstream, Chiangmai, Northern Thailand

Figure 31. Maeya Waterfall Erosion: Looking Downstream, Chiangmai, Northern Thailand

73

A **delta** is the junction of a river and a lake or sea. The water slows down here and loses the power to carry sediment, which it drops at the mouth of the river. Deltas are important to human activities as well as to fish and other wildlife because they are normally home to very fertile soil and a large amount of vegetation.

An **estuary** is where freshwater from rivers flows into the ocean, mixing with the seawater. The finer sediment particles that stay in suspension may be flushed out to sea quite quickly, but heavier particles sink to the bottom as the flow meets salt water. Estuaries and the lands surrounding them are places of transition from land to sea and from freshwater to salt water.

# 5. SOLIDS CARRIED BY STREAMS

**Stream load** is a geologic term referring to the solid matter carried by a stream (Strahler and Strahler 2006). Erosion removes mineral material from the bed and banks of the stream channel, adding it to the regular flow of water. The amount of solid load (bed and suspended) that a stream can carry (stream capacity) is measured in metric tons per day passing a given location. Stream capacity is dependent upon:

> The stream's velocity

> The amount of water flow

> The gradient of the stream bed

Stream velocity is responsible for determining the size of particles a stream can transport, as well as the way in which it carries the particles, or load (Larson and Birkland 1994). Fine sand can be moved by streams flowing as slowly as three-quarters of a mile per hour. The faster stream water moves, the larger objects it can pick up and transport. In addition, the higher the gradient (slope), the faster the flow or increase in its rate (Strahler and Strahler 2006).

The water of streams can erode in three different places: lateral erosion occurs along the sides, widening the channel. Downcutting occurs when erosion deepens a stream bed. Finally, headward erosion occurs upslope of the channel.

All material transported by a stream is known as the load. The size of those materials determines how they will be transported downstream. Stream load is divided into three types: suspended load, bed load, and dissolved load.

**Suspended load** is generally made up of fine-grained sediment (usually silt and clay), which is carried entirely within the water column. These materials are too large to be dissolved, but too small to lie on the bed of the stream (Mangelsdorf 1990). Streamflow keeps these suspended materials from settling on the stream bed. Suspended load is the result of material eroded by hydraulic action at the stream surface bordering the channel as well as erosion of the channel itself. Suspended load accounts for the largest majority of stream load (Strahler and Strahler 2006).

**Bed load** rolls slowly along the floor of the stream. It includes the largest and heaviest materials in the stream, ranging from sand and gravel to cobbles and boulders. Along the stream bed, particles are moved via traction (sliding and rolling) and saltation, which is a bounce-like movement, occurring when large particles are suspended in the stream for a short distance after which they fall to the bed, dislodging particles from the bed. The dislodged particles move downstream a short distance where they fall to the bed, again loosening bed load particles upon impact (Ritter, 2008).

The matter in **dissolved load** is invisible. It is transported in the form of chemical ions. The dissolved load consists mainly of $HCO_3^{2-}$ (bicarbonate), $Ca^{2+}$ (calcium), $SO_4^{2-}$ (sulfate), $Cl^-$ (chloride), $Na^+$ (sodium), $Mg^{2+}$ (magnesium), and $K^+$ (potassium). The load of such ions can be high when there is substantial ground water recharge. The ions are eventually carried to the oceans, giving them their salty character. This type of load can result from mineral alteration, from chemical erosion, or it may even be the result of groundwater seepage into the stream. Materials comprising the dissolved load have the smallest particle size of the three load types (Strahler and Strahler 2006).

# 6. COASTAL EROSION BY CURRENTS

Coastal erosion is the wearing away of land and the removal of beach or dune sediments by wave action, tidal currents, wave currents, or drainage (Figure 32).

Figure 32. Coastal Erosion by Currents, Cote de granite Rose, Brittany Coast near Ploumanach, France

Coastal erosion is a natural process that takes place over a range of time scales. It may occur in response to:

> Smaller-scale (short-term) events, such as waves generated by tides, winds, and storms. (The power of oceanic waves is awesome. Large storm waves can produce two thousand pounds of pressure per square foot.)

> Larger-scale (long-term) events, such as glaciations or orogenic cycles that cause coastal erosion and may significantly alter sea levels up or down, and tectonic activities that cause coastal land subsidence or emergence.

Consequently, most coastlines are naturally dynamic, and cycles of erosion are often an important feature of their ecological character.

Coastal erosion is a natural process, but it may become a hazard when humans interact with the coastal environment by developing it and creating value there. Coastal erosion is widespread in the coastal zone in the Indian Ocean owing to a combination of economic development and population growth. This has led to major efforts to manage the situation and changes induced by human activities, extreme events, and sea level rise. These efforts focus on the roles that coastal forest and trees can play in combating coastal erosion.

## 7. ACTION OF FROST AND ICE

**Frost** may act in three different ways to influence surface erosion (Satterlund and Adams 1992):

> Cohesive forces holding together a soil may be overcome by the expansion of water, causing detachment of soil particles from the surface.

> Soil frost may prevent water from infiltrating, resulting in greater overland flow.

> Soil frost may become a source of water for overland flow as it melts, even without rain or snowmelt.

Though not common in areas receiving significant snowpack (snow insulates the soil surface), soil frost is a significant erosive factor where bare or sparsely vegetated soils are rarely covered by snow and where freezing temperatures are common.

**Ice erosion**, the erosive power of moving ice, is evident where glaciers have melted away, exposing the bedrock. Ice actually has more erosive power than water, but since water is much more common, it is responsible for more erosion on Earth's surface.

Glaciers can perform two erosive functions—they pluck and abrade. Plucking, or quarrying, is when water enters ground cracks under the glacier, freezes, and breaks off pieces of rock. The eroded material is transported until it is deposited or until the glacier melts. Abrasion cuts into the rock under the glacier, scooping rock up like a bulldozer and smoothing and polishing the rock surface.

## 8. HUMAN CAUSES OF EROSION

While erosion is a natural process, human activity on and around Earth's surface has caused ten to forty times more erosion than have the natural processes. Agriculture and construction are the two major causes of erosion. Agriculture has been a primary driver of deforestation since ancient times. The vast, old-growth forests that once covered much of the world largely have been cut and burned down to make way for agriculture. Construction and road building usually involve removing existing vegetation. Any land-disturbing activity that removes plant cover and exposes the soil surface to rainfall and flowing water accelerates the erosion process.

Livestock grazing is also a human cause of erosion because it makes the ground's surface soil bare and extremely prone to erosion by natural forces.

## 9. CONCLUSION

Water plays an important role in the transport of a large amounts of soil in the form of sediment and thus in the transformation of Earth's landscape. Sediment is eroded from the landscape, transported by river systems, and eventually deposited in lakes, seas, or oceans. Natural, geologic erosion takes place slowly over centuries or millennia, whereas erosion resulting from human activities can be much faster.

The transportation process is initiated on the land surface. Frost, ice, rivers, and streams act as conduits for sediment movement. The higher the discharge, the higher the velocity and the greater the capacity for sediment transport.

Sediment is measured and classified according to its dynamic characteristics: as suspended load, as bed load, or as dissolved load.

# DEPOSITION OF SEDIMENTS

## 1. INTRODUCTION

Water is responsible for a large amount of sediment transport. Sedimentation is the direct result of the loss (erosion) of sediments from other aquatic areas or land-based areas. Sedimentation is a natural process of all water bodies (i.e., lakes, rivers, estuaries, coastal zones, and even the deep ocean). Sedimentation can be detrimental or beneficial to aquatic environments. Let's look at sediment deposition on Earth's surface and the role of water in the process.

## 2. STRATIFIED SEDIMENTARY DEPOSITS

Stratification is a fundamental feature of sedimentary rocks. It may result from changes in texture or composition during deposition; it may also refer to the way sediment layers are stacked over each other. It can occur on the scale of hundreds of meters as well on the submillimeter scale. Superposed strata in sedimentary rocks may appear as alternations of coarse and fine particles, as a series of color changes resulting from differences in mineral composition, or merely as successive layers of sediment deposited periodically with interruptions of sedimentation.

Water plays an important role in sorting sediments of different sizes, weights, and shapes of particles. Where layers have been deformed, the record of past movements of Earth's surface is preserved in the stratification.

The important and distinctive structural types of strata are characteristic of particular environments. They are:

**Cross-bedding,** which forms from running water. As water flows, it creates bedforms, such as ripples or dunes, on the floor of a channel. Cross beds are sets of beds that are inclined relative to the direction in which the water (or wind) was moving at the time of deposition and these inclined layers. Nevertheless, most layering is parallel. Each bed represents a homogeneous set of conditions of sedimentation. New beds indicate new conditions. Boundaries between sets of cross beds usually represent erosional surfaces.

**Graded beds.** A decrease in a current's velocity results in bedding showing a decrease in grain size from the bottom of the bed to the top. Therefore, the bedding has the coarsest sediment at the bottom and finest at the top. These often form in submarine canyons but are found in surface stream deposits too. A collection of graded beds is termed a turbidite deposit (indicating a sudden, strong current that deposits heavy, coarse sediments first, with finer ones following as the current weakens).

**Ripple marks** are produced by flowing water or wave action. We often see waves of sand on beaches at low tide and in stream beds.

> ❭ **Asymmetrical** ripples, often found in rivers, contain a steeper slope downstream with an alternation in current flow from one direction (Marshak 2008). Ripples preserved in ancient rocks can also indicate an up-and-down direction in the original sediment.

> ❭ **Symmetrical** ripples, often found on beaches, tend to have the same slope on both sides of their crests (Marshak 2008). Symmetric ripples form as a result of constant wave energy oscillating back and forth.

**Mud cracks** form when a water-rich mud dries out in the air. The cracks form due to shrinkage of the sediment as it dries. The presence of mud cracks indicates that the sediment was exposed at the Earth's surface and then rapidly buried. Mud cracks are common in arid environments, such as deserts.

# 3. FLUVIAL AND MARINE DEPOSITS

**Fluvial** refers to the processes associated with rivers and streams and the deposits and landforms created by them. Fluvial sediment is found where water is the key agent for erosion. When streams or rivers are associated with glaciers, ice sheets, or ice caps, the terms glaciofluvial or fluvioglacial are used (Neuendorf et al. 2005, Wilson and Moore 2003).

A river continually picks up and drops solid particles of rock and soil from its bed throughout its length. In areas of fast flow, as in a mountain river, more particles are picked up than dropped. Conversely, in slow-flow areas, more particles are dropped than are picked up. The dropped particles are alluvial deposits, which consist of silt, sand, clay, and gravel, as well as a lot of organic matter.

The amount of matter carried by a large river is enormous. It has been estimated that the Mississippi River annually carries 406 million tons of sediment to the sea (Mathur and da Cunha 2001).

**Marine-deposition coasts** are formed by accumulation of sediments through wave action (Figure 33). Marine-deposition coasts are accompanied by these principal elements:

> **Beaches.** These are the main features of coastal deposition. A beach is defined as the gently sloping area of land between the high- and low-water marks. Beaches are made up of material transported along the coast by long shore drift (such as sand and shingle). Beaches are not permanent features, as their shapes can be altered by waves every time the tide comes in and goes out. Shingle beaches have a steeper gradient than sandy beaches.

> A **spit** is a deposition landform found off coasts. It is a long, narrow ridge of shingle and sand with one end attached to the land and the other projecting at a narrow angle, either into the sea or across a river estuary. Spits are formed where the prevailing wind blows at an angle to the coastline, resulting in long shore drift.

> A **bar** may separate a lake from the sea. The grain size of the material comprising a bar is related to the size of the waves and the strength of the currents moving the material. A bar is formed in the same way as a spit initially, but a bar is a spit that continues to grow across a bay and joins two headlands, creating a lagoon behind it. The lagoon traps water, but it

may gradually be infilled as a salt marsh develops in its low-energy zone (which encourages deposition).

❯ A **tombolo** is a ridge of beach material (typically sand) built by wave action that connects an island to the mainland. Tombolos are formed where a spit continues to grow outward, joining land to an offshore island.

All of these forms consist of sand, gravel, shell detritus, or a mixture of them, and they begin similarly.

Figure 33. Marine-Deposition Coasts: Sandy Beach Mountains, Nang Yuan Island

## 4. SHOALS AND SILTING

A **shoal** or sandbar is a somewhat linear landform within or extending into a body of water, typically composed of sand, silt, or small pebbles. It is characteristically long and narrow (linear). It develops where a stream or ocean current promotes deposition of granular material, resulting in localized shallowing (shoaling) of the water. In addition, a shoal is a place where a sea, river, or other body of water is shallow (Figure 34).

Shoal deposits at the bottom of a water body are accumulations of heavy, solid materials flowing along a silty stream. The silting may be caused by soil erosion or by disposal of solid waste from industries. Rivers can be blocked by large amounts of sediment. Silt is created by a variety of physical processes capable of splitting the generally sand-sized quartz crystals of primary rocks by exploiting deficiencies in their lattice (Moss and Green 1975). Silt can result from chemical weathering of rock and regolith[20] (Nahon and Trompette 1982) and a number of physical weathering processes, such as frost shattering (Lautridou and Ozouf 1982) and haloclasty[21] (Goudie and Viles 1995).

Shoals are formed not only by direct sedimentation of solid materials but also by the movement of settled materials during bed erosion (Shaheen and Chantarasorn 1971).

Runoff is very slow in shoal soils, and permeability is moderate. Available capacity of water is high. Shoal soils are used primarily for cultivated crops or for pasture.

## 5. DEPOSITS OF SALINE RESIDUES

Rock salt is typically formed by the evaporation of salty water (such as seawater). Rock salt is the common name for halite. Its chemical formula is NaCl (sodium chloride), a chemical compound belonging to the larger class of ionic salts.

---

20  Regolith is a layer of loose, heterogeneous material covering solid rock.
21  Haloclasty is a physical weathering caused by the growth of salt crystals.

Figure 34. Amanohashidate Sandbar in Miyazu Bay, Northern Kyoto Prefecture, Japan

Sodium chloride occurs naturally in large quantities. Rock salt can be found all over the world. Saline residues, or salts, originate from the natural weathering of minerals or from fossil salt deposits left from ancient seabeds. Salt deposits can ring dry lake beds, inland marginal seas, and enclosed bays and estuaries in arid regions of the world. The oceans are by far the largest storehouses of salt. Several billion years ago, very large bodies of water evaporated or shrank, making enormous deposits of rock salt (such as that found in the Mediterranean Sea). Underground deposits of solid rock salt are also of marine origin.

Rock salt is essential for life, but it can be harmful to animals and plants in excess. Salt and saltiness is one of the basic human tastes. Rock salt is also applied to roadbeds and other impervious surfaces as a deicer.

## 6. DEPOSITS FROM HOT SPRINGS, GEYSERS, AND FUMAROLES

Hot springs, geysers, and fumaroles are all geothermal features. Surface water percolates downward through the rocks below Earth's surface to high-temperature regions. There, the water is heated and then rises to the surface in the form of a geothermal feature.

**Hot springs** occur when rainwater seeps and percolates to a certain depth and earth temperature. Some springs are thousands of years old. The earth's pressure causes the water to rise above the surface, and it is heated by geothermal heat. The geothermal gradient refers to the temperature of rocks within the planet, increasing with depth. The heated water rises back up to the surface rapidly without time to cool, creating a natural hot spring.

Hot springs vary in temperature and can be calm, effervescent, or boiling. If water percolates deeply into the crust, it heats when it contacts hot rocks, and during its travels up, it dissolves material from the surrounding bedrock and brings it up to the surface. Thus, hot springs tend to be full of minerals that people have used for medicinal purposes. These types of hot springs are produced where water is heated in nonvolcanic areas. However, not all hot springs are safe for bathing. Some are much too hot and/or much too acidic and can severely injure anyone setting foot in them.

**A geyser** is a rare kind of hot spring that is under pressure and erupts, sending jets of water and steam into the air. Unlike hot springs, where the heated water has a simple path up to the surface, geysers come from tubelike holes that run deep into the crust. The tubes fill with water, and near the bottom is molten rock called magma, which heats the water. The water down near the magma becomes superhot, but due to the immense pressure at that depth, the water cannot boil. The pressure decreases as the superheated water rises to the surface and gradually, the water starts to boil. As some of it is forced upward, it steams (turns to gas). This release of steam allows some of the water to overflow, out of the geyser's mouth. The powerful jet of steam ejects the column of water above it as water rushes up through the tube and into the air. The eruption continues until all the water is forced out of the tube or until the temperature inside the geyser drops below boiling (100°C). Groundwater seeps back into this underground network over time, starting the cycle all over again.

**A fumarole** is a hot spring that boils off all of its water before the water reaches the surface. It can be a hole in the flank of an active volcano or in a geothermal field, where temperatures are generally close to the boiling point of water. Fumaroles vent steam and gases, such as carbon dioxide, sulfur dioxide, hydrochloric acid, and hydrogen sulfide. The steam is created as pressure drops on the superheated water that emerges from the ground. Fumaroles may occur along tiny cracks or long fissures, in chaotic clusters or fields, and on the surfaces of lava flows and thick deposits of pyroclastic flows. Depending on the heat source, fumaroles can be short- or long-term features. They may persist for decades or centuries if they are above a persistent heat source. They may disappear within weeks or months if they occur atop a fresh volcanic deposit that cools quickly.

Hot springs are more common than geysers and fumaroles. Geysers are hotter than hot springs, with periodic eruptions of hot water and steam through a vent on the surface of the planet. Fumaroles are hotter than hot springs and geysers. Any groundwater that enters a fumarole is instantly turned into steam—no liquid water is present in fumaroles.

# 7. CONCLUSION

Deposition is the laying down of sediment carried mainly by water. When there is not enough energy to transport the sediment, it comes to rest. Waterways move many millions of tons of sediment annually in this never-ending cycle of erosion, transportation, and deposition. Sediment can be transported as pebbles, sand, mud, or as salts dissolved in water. Later, evaporation may deposit salts.

# Part III:

# WATER SUPPLY ENGINEERING

# SOURCES OF WATER SUPPLY

## 1. INTRODUCTION

**Subsurface water** or **groundwater** is an important source of water supply. People receive their water mainly from groundwater and surface water. However, **surface freshwater** is unfortunately limited and unequally distributed around the world. Although millions of lakes are scattered over Earth's surface, most are located in higher latitudes and mountainous areas. Almost 50 percent of the world's lakes are located in Canada alone (UNEP 2002). Let's look at some basic information about the sources of drinking water typically used in the home.

## 2. SUBSURFACE WATER

Subsurface water comes from beneath the surface of ground with soil pores or fractures in rock. We can obtain it from infiltration galleries, infiltration wells, and springs, for example.

### ■ 2.1. INFILTRATION GALLERIES

An infiltration gallery is an artificial tunnel that extends into the zone of saturation. Water flows by gravity from the zone of saturation to the land surface or into a sump or well (Meinzer 1923). Galleries are often used in conjunction with other water supply systems to increase water intake in areas of poor water yield. Infiltration galleries are permeable horizontal or inclined conduits (pipes) into which water can infiltrate from an overlying or adjacent source. They are constructed below the water table, where there is sufficient recharge to offset the pumping rate and where the permeability of the soil is sufficient to transmit enough water to the gallery under the existing head conditions (Singh Jurel et al. 2013). The galleries are usually surrounded with a gravel pack to improve flow toward it and to filter any large particles that might block its perforations.

Infiltration galleries may be used to produce drinking water through the per-forated pipes. These are laid about five to ten meters below a riverbed and connected to a sump well into which water, filtered through the sandy river-bed, flows under gravity. The stored water is treated for drinking safety. Even when there is negligible surface flow in the river, there is always subsurface flow in the permeable strata of sand below its bed, which filters water con-tinuously (Singh Jurel et al. 2013).

Infiltration galleries vary in size. Some are a few meters long and feed into a spring box. Others are many kilometers long and form an integral part of an urban water supply.

## ■ 2.2. INFILTRATION WELLS

An infiltration well is an excavation or structure created in the ground by digging, driving, boring, or drilling to access groundwater in underground aquifers. Pumping a well lowers the water level in it, which, in turn, forces water to flow in from the aquifer. Thin or impermeable aquifers may need several times as much drawdown (lowering) for the same yield as perme-able ones need, and they frequently yield only small supplies. Thick or per-meable aquifers may yield several million gallons daily with a drawdown of only a few feet.

A water supply well is an artificial excavation for getting water for drinking or other domestic use, constructed by any method. Wells can vary greatly in depth, water volume, and water quality. Well water typically contains more minerals in solution (that is, it is hard water) than surface water contains and may require treatment to soften it.

**Driven wells** are still common today. They are built by driving a small-diameter pipe into soft earth, such as sand or gravel.

**Dug wells** are holes dug by shovel or backhoe. Dug wells several feet in diameter are frequently used to reach shallow aquifers. They have been used as a household water supply source for many years. Most existing dug wells are relics of older homes. They were dug before drilling equipment was readily available or when drilling was considered too expensive.

**Drilled wells** can be more than a thousand feet deep. Often, a pump is placed at the bottom to push water up to the surface.

The distance between wells must be sufficient to avoid harmful interference when the wells are pumped. In general, the distance between two wells varies directly with the quantity of water to be pumped and is determined by the porosity of the material from which the water is drawn. The finer the water-bearing material, the farther apart the wells should be. Wells may be a few feet apart to a mile or more apart.

## ■ 2.3. SPRINGS

A spring is a place where groundwater emerges naturally. The water source of most springs is rainfall that seeps into the ground uphill from the spring outlet. Springs generally supply small quantities of water and hence are suitable for small hill towns. Some springs discharge hot water due to presence of sulfur and are useful only to cure certain skin diseases. Because of their variability, springs need to be selected with care, developed properly, and tested periodically for contaminants. There are many types of springs:

> **Gravity springs** occur in unconfined aquifers. Where the ground surface dips below the water table, the depression will fill with water.

> **Surface springs** are formed when an impervious stratum that supports the ground water reservoir outcrops.

> **Artesian springs** issue water under artesian pressure, generally through some fissure or other opening in a confining bed that overlies its aquifer. They are also known as fissure springs.

# 3. SURFACE WATER

Water from a source must be transmitted to the community or area to be served and distributed to individual customers. The major supply conduits or artificial channels for conducting water from a distance (from the source to the distribution system), usually by means of gravity, are called mains or aqueducts.

Natural sources, such as rivers, lakes, and impounding reservoirs, can be sources of surface water. Surface water supply is one of the most important hydrologic parameters for agricultural water management and drought and flood alarms.

Water is withdrawn from rivers, lakes, and reservoirs through intakes. Water-intake structures divert water from a river or channel for municipal water supply, hydroelectric power, and irrigation. Historically, water diversion has relied on large-scale structures, the engineering of which can be complicated. Depending upon the source of water, intake works can be river intakes, lake intakes, canal intakes, and floating intakes.

**River intakes.** In some areas, water is taken directly from a river into the water supply system. When mountain streams are tapped well above permanent human habitation and where the sediment content and bed load transport are low, water may only require filtration and terminal disinfection before it is used. A river intake consists of a port (conduit) provided with a grating and a sump or gravity well. The water is abstracted through a screen over a canal (usually made of concrete and built into the riverbed). River intakes are likely to need screens to exclude large floating matter. The bottom of the river intake must be sufficiently stable. The bars of the screen are laid in the direction of the current and slope downward so that coarse material cannot enter. From the canal, water enters a sand trap and then may pass a valve and flow by gravity, or be pumped into the rest of the system (Jordan 1984, Lauterjung and Schmidt 1989).

**Lake intakes.** Water can be abstracted for drinking water supplies from either natural lakes or artificial ones (created by dams). The quality of water in lakes varies widely. In small lakes and ponds, there is a high risk of fecal pollution.

Lake intakes are usually constructed in the bed of the lake, below the water, so that they can draw water in dry seasons also. These intakes are no obstruction to navigation and are in no danger from floating bodies or ice. Because lake intakes draw small quantities of water, they are not used in big water supply schemes, on rivers, or in reservoirs, mostly because they are not easily approachable for maintenance (Rao 2005).

In deep lakes, the intake should be some meters below the surface and in some cases deeper, usually because of excessive algal growth on the lake surface. In shallow lakes, the intake should be sufficiently high above the lake floor to prevent the entrance of silt.

**Canal intakes** are used where they can be built economically to follow the hydraulic gradient (slope of the flow). If the soil is suitable, the canals are excavated with sloping sides and are not lined. Otherwise, concrete or asphalt linings are used. Canals run across obstructions like rivers, streams, natural drains, and other canals at different points; these crossings with such obstacles cannot be avoided. The various structures at the crossing points must be constructed for the easy flow of canal water and drainage. These structures are known as cross drainage works.

A fine screen is provided over the bell mouth of a canal intake's outlet pipe. The bell mouth entry is located below the expected low water level in the canal. Water may flow from outlet pipe under gravity if the filter house is situated at a lower elevation. Otherwise, the outlet pipe may serve as suction pipe, and the pump house may be located on or near the canal bank (Punmia et al 1995). An intake chamber is constructed in the canal section. This results in the reduction of the waterway, which increases the velocity of flow. It therefore becomes necessary to provide pitching on the downstream and upstream portions of a canal intake (Rao 2005).

A **floating intake** is a drinking water system that allows water to be taken directly from the dug-out through a pipe, thus avoiding the problems encountered in infiltration trenches.[22] The inlet pipe of a suction pump is connected just under the water level to a floating pontoon moored to the bank or bottom of a lake or river. The pump itself can be located either on the bank or on the pontoon. The advantage of the pontoon pump is that the suction pipe can be quite short, and the suction head can be constant (Hofkes and Visscher 1986).

## 4. CONCLUSION

Sources of water supply are classified as subsurface water and surface water. Water is obtained from subsurface sources through infiltration galleries, infiltration wells, and springs. Infiltration galleries are constructed below the riverbed to draw water, particularly outside of monsoon season. Infiltration wells include driven wells, dug wells, and deep wells. They

---

22  Infiltration trenches are shallow excavations lined with rubble or stone that are designed to infiltrate storm water though permeable soils into the groundwater aquifer.

supply water from more than one water-bearing stratum. Springs come in many types, such as gravity springs and artesian springs. They supply small quantities of water, typically for drinking. Surface water includes rainfall, lakes, ponds, rivers, and reservoirs. It is withdrawn by constructing intake structures, which are classified as river intakes, lake intakes, canal intakes, and floating intakes.

# WATER QUALITY

## 1. INTRODUCTION

Freshwater resources provide essential services to society. The most important of these services is fresh drinking water for municipalities around the world. Other freshwater resources provide other services, such as power generation and water for recreational activity. However, pollution of these freshwaters can reduce the water quality enough to make it unsafe for drinking or other domestic uses.

Water quality determines the state of aquatic environments. It is the combined chemical, physical, and biological characteristics of water (Diersing 2009). It is a measure of the condition of water relative to the requirements of one or more biotic species and/or to any human need or purpose (Johnson, et al. 1997).

Let's look at the most common scientific measurements used to assess water quality for drinking, safety of human contact, and the health of ecosystems.

## 2. WATER CONSUMPTION

The most obvious ways we use water are for drinking, cooking, bathing, cleaning, and—for some—watering gardens for growing food. Many businesses, notably the food and beverage and pharmaceutical sectors, consume water by using it as an ingredient in finished products for human consumption.

Industry uses water to make steam for direct-drive power and for use in various production processes or chemical reactions, such as for metals, wood, paper, chemicals, gasoline, and oils. Industrial reliance on water makes it essential that we preserve water in every way possible and make sure water pollution is kept at minimal levels.

Industrial use of water increases with a country's income, going from 10 percent for low- and middle-income countries to 59 percent for high-income countries (UNE and UNESCO 2003). The largest single use of water by industry is for cooling in thermal-power generation by thermoelectric power plants (whether the fuel is coal, oil, gas, nuclear fuel, or biomass).

Contaminants that may be found in untreated water include bacteria, nitrates, metals, trace quantities of toxic materials, and salts. The quality of water depends on the geology and hydrology of the area in which it lies, as well as human activities, including coastal development, agriculture, aquaculture, mining and industry, and shipping. The Safe Drinking Water Act (SDWA) authorizes the US Environmental Protection Agency (EPA) to issue two types of standards: primary standards regulate substances that potentially affect human health, and secondary standards prescribe aesthetic qualities—those that affect taste, odor, or appearance.

Countries need governing bodies to control water use and to establish the right plan of water management. Freshwater is needed for industries to survive, but it is also needed for humans to survive. We need to follow conservation plans to make freshwater accessible to all.

# 3. SAMPLING AND MEASUREMENT

The complexity of the subject of water quality is reflected in the many types of measurements of the physical and chemical conditions of water. The most accurate measurements of water quality are made on site and in direct contact with the water (temperature, pH, dissolved oxygen, conductivity, turbidity, and so on). Because water exists in equilibrium with its surroundings, scientists collect water samples, take photographs from airplanes and even satellites, and observe what's happening along streams, lakes, and bays to get an overall sense of the health of the water.

The **temperature** of water can affect it in many different ways. Some organisms prefer cool water, while some like it warm. Most aquatic organisms are cold-blooded. This means that the temperature of their bodies matches the temperature of their surroundings. Scientists measure water temperature for several reasons. First, it determines the kinds of animals that can survive in a given body of water. Temperature affects organisms' photosynthesis and digestion, and if the temperature gets too hot or too cold for some of them, they die. Temperature also can affect the chemistry of the water. For example, when its temperature goes up, water holds more dissolved solids (like salt or sugar) but fewer dissolved gases (like oxygen). The opposite is true for colder water. A healthy cluster of trees and vegetation next to a stream or river helps keep temperatures cool for trout and other fish. Plants and algae that use photosynthesis prefer to live in warm water, where there is less dissolved oxygen. Generally, bacteria tend to grow more rapidly in warm waters.

**Oxygen** is necessary for many aquatic species to survive. Scientists measure dissolved oxygen (DO) because it tells them how well fish and other aquatic organisms can breathe. Most healthy water bodies have high levels of DO. Certain water bodies, like swamps, naturally have low levels of DO. Several factors can affect how much DO is in the water. These include temperature, the amount and speed of flowing water, the plants and algae that produce oxygen during the day and take it back in at night, pollution in the water, and the composition of the stream bottom.

The **pH** is a measure of acidity, as we have seen. It ranges from 0 (extremely acidic) to 14 (extremely basic) with 7 being neutral. Scientists measure pH to determine the concentration of hydrogen in the water. Most waters range from 6.5 to 8.5. Substances that are close to pH 7 work well with our bodies; the same holds true for aquatic organisms. Changes in pH can affect how chemicals dissolve in the water and whether they affect organisms. If the water becomes too acidic or basic, it can kill. High acidity can be deadly to fish and other aquatic organisms. However, not all acids and bases are bad. Aspirin and tomatoes are acidic, while milk of magnesia and baking soda are both bases.

**Turbidity** refers to the clarity of water (how clear it is). This determines how much light gets into the water and how deep it penetrates. Scientists measure the clarity of water to determine how many particulates are floating around. Excess soil erosion, dissolved solids, or excess growth of microorganisms can cause turbidity, and the sunlight will not be able to penetrate very far. All plants need sunlight to grow, and in dirty water, there may not be enough light to support plant growth. When its plants die, there is less dissolved oxygen in water. Dead plants also increase the organic debris, which microorganisms feed on, and thus the number of decomposers increases. These decomposers grow rapidly, using up a great deal of oxygen, leading to further depletion of oxygen. If there is no dissolved oxygen in water, other aquatic life-forms cannot live in it.

Scientists test all of these parameters relative to the time of year. If you're sitting on a dock in a pond on a warm summer day, you might be able to see to the bottom, which means the pond has low turbidity. But after a rainstorm, when all the muck has been stirred up, you won't be able to see the bottom: the pond then has high turbidity. Scientists use turbidity measurements to calculate the inputs from erosion and nutrients.

**Nutrients** in water are critical for plants and animals. The two major nutrients that scientists measure are nitrogen and phosphorus. Nitrogen is essential to the production of plant and animal tissue. It is required in large quantities for synthesis of proteins and nucleic acids. Nitrogen enters the ecosystem in several chemical forms and occurs in other dissolved or particulate forms, such as tissues of living and dead organisms. Phosphorus is also essential for life. As phosphate, it is a component of DNA, RNA, ATP, and the phospholipids that form all cell membranes. Phosphorus is rarely found in high concentrations in freshwater, as plants actively take it up. However, the presence of too many nutrients can hurt aquatic organisms by causing lots of algae to grow in the water. Nutrients can also affect pH, water clarity, and temperature, and they can cause water to smell and look bad.

**Nitrate**, a compound containing nitrogen, can exist in the atmosphere or as a dissolved gas in water. At elevated levels, it can have harmful effects on humans and animals. Nitrates in drinking water can be a cofactor in causing methemoglobinemia in infants (Fewtrell 2004). Common sources of excess nitrate in lakes and streams include septic systems, animal feedlots, agricultural fertilizers, manure, industrial wastewaters, sanitary landfills, and garbage dumps. Excess nitrate can make it difficult for aquatic insects and fish to survive.

The presence of **toxic substances** in water may have several effects. One of the main concerns is bioaccumulation, or the "accumulation of persistent chemicals in the tissues of living organisms" (NSCEP). Bioaccumulation of heavy metals is known or suspected to cause cancer, birth defects, reproduction problems, and other serious illnesses. Exposure to certain levels of some toxic substances can cause difficulty in breathing, skin rash, headache, nausea, or other illnesses. Certain toxic pollutants in drinking water can even cause death.

Scientists test for many harmful substances in water, such as metals, pesticides, and oil. For example, mercury can limit the amount of dissolved oxygen in water. When fish have less oxygen available for respiration, they die (especially in lakes and estuaries). Mercury comes from mining, natural sources, and air pollution from power plants and incinerators. People are warned not to fish in streams, lakes, or bays where high levels of mercury or other harmful substances are found.

Not all measurements are chemical or physical. **Visual surveys** can reveal observable problems on a stream, for instance, and help characterize the environment through which it flows. Scientists take measurements of the landscape surrounding a stream to determine how many trees and shrubs grow along it, the amount of shade they create, and how much woody debris (sticks and leaves) is in the stream. Vegetation, tree cover, and woody debris create habitats for wildlife and fish. Vegetation can filter pollution from the runoff flowing through it, reducing sediment and other solids and particulates as well as associated pollutants, such as hydrocarbons, heavy metals, and excess nutrients.

Scientists sample water for certain types of **bacteria** that are found only in the intestinal tracts of humans and other warm-blooded animals (such as pets, livestock, and wildlife). Most **bacteria** are not harmful and do not cause human health problems. The bacteria that scientists measure, which are called fecal coliforms, are not necessarily harmful, but they usually hang out with certain viruses and pathogens that can make you sick. Escherichia coli (E. coli) is one subgroup of fecal coliform bacteria. E. coli bacteria are good indicator organisms of fecal contamination. The major sources of fecal coliforms in water are failing septic systems, wastewater treatment plants, boat discharges, and animal waste (which covers a big range, from pet droppings to cow manure).

**Biological sampling** aims to distinguish between naturally occurring variation and changes caused by human activities. Scientists determine the health of waters by taking samples of fish, macro invertebrates, or plant communities to assess the health of rivers and wetlands. These organisms are found in shallow waters that are accessible either by wading or from a bank or boat. Some love to live in dirty water, so if scientists find a lot of these in a sample, they know there's a problem. Other organisms can survive only in water that's very clean, so finding those means the water is probably healthy.

## 4. CONCLUSION

Scientists use many different sampling and measurement methods to determine the quality of water. They assess physical, chemical, and biological attributes. They are important for determining the variability and the temporal dynamics of water quality, which can help separate natural influences from those of human activities. Good water quality helps produce healthy stream communities. Poor water quality can cause health, growth, or survival problems for stream life. Levels of turbidity, pH, dissolved oxygen, temperature, alkalinity, and other water-quality elements can directly affect the health of individual organisms and aquatic populations.

Achieving good water quality for drinking or for domestic use depends on a water treatment system that accurately measures and monitors contaminants. Producing pure water is only part of the equation; validating quality, storing water, and maintenance are also critical to ensuring the required water quality. Next, we'll look at some attributes of the water treatment system.

# TREATMENT OF WATER

## 1. INTRODUCTION

Water treatment describes those industrial-scale processes used to render water suitable for human use and consumption, industry, medical, and many other purposes. Such processes may be contrasted with small-scale water sterilization practiced by campers and other people in wilderness areas. The goal is to produce physically, chemically, and biologically safe water. Let's look at processes for drinking water treatment and industrial water treatment.

## 2. DRINKING WATER TREATMENT

Humans need water for drinking, sanitation, and agriculture. Many people do not have access to safe drinking water. Many serious diseases, such as cholera, are caused by drinking water that contains parasitic microorganisms. Water containing large amounts of industrial waste or agricultural chemicals (such as pesticides) can also be toxic and unfit for drinking. Hence, we need to use various methods of water treatment to provide safe drinking water for communities. Today, the most common steps in water treatment for community water systems (mainly surface water treatment) include coagulation and flocculation, sedimentation, filtration, and disinfection.

**Coagulation and flocculation** are often the first steps in water treatment. Liquid aluminum sulfate (alum) and/or a polymer are added to untreated (raw) water. There, the positive charge of these chemicals neutralizes the negative charge of dirt and other dissolved particles in the water. Next, groups of dirt particles stick together to form larger, heavier particles called **flocs** that are easier to remove by settling or filtration.

As the water and the floc particles progress through the treatment process, they move into **sedimentation** basins where the water moves slowly, causing the heavy floc particles to settle to the bottom.

Once the floc has settled to the bottom of the water supply, the clear water on top will pass through filters made of layers of sand and gravel, and in some cases, crushed anthracite. This **filtration** removes dissolved particles, such as dust, parasites, bacteria, viruses, and chemicals. The filters are routinely cleaned by backwashing.

After the water has been filtered, a disinfectant may be added to kill any remaining parasites, bacteria, and viruses, and to protect the water from germs when it is piped to homes and businesses. Chlorine is very effective at **disinfection**, and residual concentrations can be maintained to guard against possible biological contamination in the water distribution system.

# 3. INDUSTRIAL WATER TREATMENT

In industrial water treatment, heating and cooling play important roles. After agriculture, industrial water treatment is the second-largest water demand. Besides applications in processing, washing, drinking, and fire fighting, the main application of water in industry is for heat transfer in boilers and cooling towers.

Industrial water treatment is of three main types:

- ❯ Industrial wastewater treatment

- ❯ Boiler water treatment

- ❯ Cooling water treatment

### ■ 3.1. INDUSTRIAL WASTEWATER TREATMENT

Wastewater treatment refers to the process of removing pollutants from water previously employed for industrial, agricultural, or municipal uses. Disposal options include discharge to a public sewer or on-site treatment prior to discharge to a sewer or watercourse. Each of the techniques outlined below is applied through the many stages of wastewater treatment.

### ■ 3.1.1. PRIMARY TREATMENT

Preliminary treatment of wastewater includes screening and removing grit, oil, and grease.

**Screening** is the first step. Mechanical screens remove all sorts of refuse that has arrived with the wastewater (plastic, branches, rags, toilet paper residues, and metals). The screening process mainly helps prevent clogging and interference in the treatment processes that follow it. The material retained (called "screenings") is usually washed to remove fecal matter and then compressed for disposal in a landfill or incinerator.

**Grit removal** is the next primary stage. Pretreatment water may include fine mineral matter (grit and sand) originating mainly from road runoff. It is allowed to deposit in long channels or circular traps. The retained solids are removed because they may damage pumps and other equipment. They are usually sent to a landfill.

**Removal of oil and grease** is necessary at some treatment works to protect the next processes. We should not pour materials such as oil and grease down drains or discharge them to sewers.

**Sedimentation** is the last step in primary treatment: the larger solids are removed to facilitate the efficiency of the procedures that follow it and to reduce the biological oxygen demand of the water. A sedimentation tank (also called a settling tank or a clarifier) allows suspended particles to settle out of water or wastewater as it slowly flows through. A layer of accumulated solids (sludge) settles, and floating materials, such as grease and oils, rise to the surface and are skimmed off. Primary settling tanks are usually equipped with mechanical scrapers that continually drive the collected sludge toward the bottom of the tank, where it can be pumped away to further sludge treatment stages. Chemicals known as coagulants and flocculants are used to produce pinpoint (very small) floc, which is filtered from the water.

## ■ 3.1.2. SECONDARY TREATMENT

Secondary wastewater treatment (biological treatment) consists of removing or reducing the dissolved organic matter and colloidal organic matter that escapes primary treatment with the use of aerobic processes.

Wastewater is full of organic matter and nutrients. Bacteria can break it down, producing energy and building material for cell growth (carbon dioxide, water, and humus), a process similar to the way humans gain energy and material for growth from food.

**Wastewater treatment plants** try to optimize this ability of bacteria to "eat" the organics in wastewater by providing ideal conditions for their growth and metabolism. This is an environment with the proper pH, temperature, micro- and macronutrients, and oxygen levels. Quickly and effectively, bacteria then break down pollutants into less harmful components. After the major solids are removed from wastewater in primary clarification, the wastewater is sent to large pools, such as aeration lagoons or oxidation ditches. In the aeration zone, some organic matter will be used to grow new bacteria, and some will be oxidized and released as carbon dioxide. The mass of bacterial cells generated is called activated sludge.

After the aeration zone process, the mixed liquor is drained into the secondary clarifiers, where the biological sludge settles in the tank and is returned to the anaerobic zone for further treatment. (*Anaerobic* means "without oxygen.") Anaerobic bacteria use organic matter in the absence of oxygen and produce biogas consisting of methane, carbon dioxide, and traces of other gases. This process further breaks down the bacterial cell mass produced during secondary treatment and reduces its volume. The treated sludge often is applied to farmland as a form of fertilizer in a manner similar to the practice of using horse manure in a household garden.

The wastewater is then passed through a **secondary clarifier**, which performs yet another round of **sedimentation**.

The **disinfection** of wastewater with chemicals, such as chlorine, is typically the final step in wastewater treatment. Chlorination is by far the most common method of wastewater disinfection worldwide before the water is discharged into streams, rivers, or oceans (Stover et al. 1986, White 1978, WPCF 1984). Chlorine is effective in destroying a variety of bacteria, viruses, and protozoa, including *Salmonella*, *Shigella*, and *Vibrio cholera*. Although a significant portion of the pathogens is inactivated, it is difficult to identify individual pathogens within wastewater. Therefore, indicator pathogens, such as fecal coliform (E. coli), are used.

**Coagulation** and **flocculation** remove turbidity, which we have seen is cloudiness in water caused by small, suspended particles). In addition, coagulation and flocculation remove many suspended bacteria and can be used to remove color, as well.

In the flash mix chamber, chemicals are added to the water and mixed violently for less than a minute. Primary coagulants cause particles to destabilize and clump together (such as metallic salts like aluminum sulfate, ferric sulfate, and ferric chloride, and cationic polymers). Coagulant aids and enhanced coagulants can add density to slow-settling floc and help maintain floc formation. Organic polymers, such as polyaluminum hydroxychloride (PACl), are typically used to enhance coagulation in combination with a primary coagulant. Then, in the flocculation basin, the water is gently stirred for thirty to forty-five minutes to give the chemicals time to act and to promote floc formation. The floc then settles out in the sedimentation basin.

### ■ 3.1.3. TERTIARY TREATMENT

Tertiary treatment removes specific wastewater constituents that could not be removed in earlier processes, such as nitrogen, phosphorus, additional suspended solids, refractory organics, heavy metals, and dissolved solids. Tertiary treatment technologies can be extensions of conventional secondary biological treatment. They further stabilize oxygen-demanding substances in the wastewater or remove nitrogen and phosphorus. Tertiary treatment can involve technologies such as membrane filtration and constructed wetlands.

**Membrane filtration** can be either micro- and ultrafiltration or reverse osmosis (RO), also known as hyperfiltration. Although categorized as different technologies, these types of membrane filtration have a great deal in common: a thin layer of a very porous polymer (plastic) is laid onto a backing material. Membranes filter out larger particles; separation of solid-liquid solutions is done by microfiltration, and ultrafiltration separates suspended solids, colloids, bacteria, and viruses through membrane pores of between one and one hundred nanometers.

Nanofiltration is mainly used to remove ions of value (such as those of salts) and larger mono ions, such as those of heavy metals. Reverse osmosis filters separate contaminating ions from water molecules. These filters are operated under high pressure in cartridges integrated into water purification systems. Their pore sizes can be less than ten angstroms.

**Constructed wetlands** can successfully reduce nitrogen content, filter out solids, and reduce the presence of heavy metals. The first experiments using wetland macrophytes for wastewater treatment were carried out in Germany in the early 1950s. Since then, constructed wetlands have evolved into a reliable treatment technology for various types of wastewater, including sewage, industrial and agricultural wastewaters, landfill leachate, and storm water run-off (Vymazal 2010).

Physical, chemical, and biological processes combine in wetlands to remove contaminants from wastewater. The type and amount of pollutant removed depends upon the species and oxygen affinity of the organisms present in the wetland. Although constructed wetlands tend to take up a great deal of space, they can be created at lower costs than other treatment options and with low-technology methods where no new or complex technological tools are needed.

## ■ 3.2. BOILER WATER TREATMENT

Boiler water treatment protects boiler and steam systems. It requires advanced water technologies and advanced formulas of corrosion and scale inhibitors. Four tools can be used in boiler water treatment: internal treatment, reverse osmosis, electrodialysis, and deaeration.

**Internal treatment** is conditioning the boiler water with chemicals. They can react with the hardness of incoming water to prevent it from precipitating onto the boiler's metal as scale, condition any suspended matter (such as hardness sludge or iron oxide) so that it doesn't stick to the boiler, eliminate oxygen from the feedwater to provide enough alkalinity to prevent boiler corrosion, and control the causes of boiler water carryover.[23]

---

23  Carryover is any contaminant that leaves the boiler with the steam. It can be in a solid, liquid, or vapor form.

Chemicals used in this process may include:

» Phosphate and polyphosphate dispersants

» Natural and synthetic dispersants

» Sequestering agents

» Oxygen scavengers

» Anti-foaming agents

In addition, an internal treatment should prevent corrosion and scaling of the feedwater system and protect against corrosion in the steam condensate systems.

**Reverse osmosis** water purification systems are increasingly employed across many industries, and the capabilities of these systems to clean water are impressive. RO filters water through semipermeable membranes. RO is growing as a method to serve high-pressure boiler feed systems. It significantly reduces salt and most other inorganic material in the water, plus some organic compounds. With a quality carbon filter to remove any organic materials that get through the filter, the purity of the treated water approaches that produced by distillation. RO units usually remove microscopic parasites (except viruses). RO systems, which do not use electricity, can typically purify more water per day than distillers do and are less expensive to operate and maintain.

**Electrodialysis** is the demineralization of water and other fluids containing ionic-form impurities (transport salt ions) from one solution through ion-exchange membranes to another solution under the influence of an applied electric potential difference.

Electrodialysis is different from reverse osmosis in that dissolved chemical species are moved away from the feed stream, rather than the reverse. Because the quantity of dissolved species in a feed stream is far lower than that of the fluid, electrodialysis offers the practical advantage of much higher feed recovery in many applications (Davis 1990, Strathmann 1992, Mulder 1996, Sata 2004, Strathmann 2004).

**Deaeration** is nearly complete oxygen removal, which is done to meet industrial standards for oxygen content and the allowable metal oxide levels in feedwater. This can be accomplished only by efficient mechanical deaeration supplemented by a properly controlled oxygen scavenger, which is used in the upstream oil and gas industry to remove oxygen from seawater, preventing or reducing corrosion and the growth of bacteria.

Vacuum deaerators can be used when heating of water is not necessary and when steam is not available as a stripping medium. Membrane contractors are increasingly being used. Carbon dioxide is also often removed using a physical medium.

## ■ 3.3. COOLING WATER TREATMENT

Cooling water treatment protects systems against damaging corrosion. It also controls scale formation and fouling that can impair cooling efficiency. It also controls the growth of harmful microbes, such as Legionella bacteria. There are three types of cooling water treatment: corrosion inhibitors, scale/deposit controllers, and microbiocides.

## ■ 3.3.1. CORROSION INHIBITORS

A corrosion inhibitor is a chemical substance that, when added in a small concentration to an environment, effectively minimizes or prevents the corrosion rate of a metal exposed to that environment (Nathan 1973). Corrosion is the degradation of any material, including polymers, ceramics, and composites. Liquid-phase inhibitors are classified as anodic, cathodic, or mixed inhibitors, depending on whether they inhibit the anodic, cathodic, or both electrochemical reactions.

Anodic inhibitors typically are used in near-neutral solutions where sparingly soluble corrosion products, such as oxides, hydroxides, or salts, are formed (Papavinasam 2000).

Cathodic inhibitors reduce corrosion primarily by interfering with or preventing the oxygen-reduction reaction.

The choice of a suitable inhibitor depends on the cooling system design parameters, the water composition, the type of metals in the system, the stress conditions, the water velocity, the pH, the amount of dissolved oxygen and salt, and the suspended matter composition. The most common inhibitors of this category are the silicates and the phosphates.

### ■ 3.3.2. SCALE/DEPOSIT CONTROLLERS

Deposition of scale is a chemical process that results when the concentration of dissolved salts in the cooling water exceeds their solubility limits, and salts precipitate on surfaces in contact with the water. Deposit accumulations in cooling water systems act as a thermal insulator to reduce heat transfer efficiency in production equipment, which increases energy consumption and costs. In addition, the deposits lead to the creation of oxygen-differential cells, which will accelerate corrosion and lead to process equipment failure. Typical components of scale found in cooling water systems include calcium salts, zinc phosphate, silica, and magnesium silicate. These scale components exhibit reverse solubility in that they become less soluble as pH and water temperature increase, and then they tend to deposit and adhere as scale on heat-transfer surfaces with higher temperature.

Since the thermal conductivity of scale is substantially less than that of metal, the scale adhesion lowers the thermal efficiency of the heat exchangers. Agents that control cooling water deposit function via crystal-growth distortion, dispersion, sequestration, and other inhibitor methods.

Scale can be controlled or eliminated with measures such as:

> Controlling cycles at a set level

> Chemical scale inhibitor treatment

> pH adjustment with acid addition

> Softening of cooling water system makeup

### ■ 3.3.3. MICROBIOCIDES

Cooling tower microbiocides are chemical compounds added to cooling tower water to prevent fouling of the system with bacterial growth. Microbiocides inhibit microorganisms in a variety of ways. Some alter the permeability of cell walls, interfering with vital life processes. Heavy metals enter the cytoplasm by penetrating the plasmalemma, destroying protein groups essential for life. Microbiocides can eradicate sulfate-reducing bacteria, aerobic, and anaerobic bacteria.

Microbiocides most commonly used for cooling water treatment are usually classified into two groups: oxidizing and nonoxidizing. Oxidizing biocides chemically oxidize the cellular structure of an organism, killing it. Organisms cannot develop immunity to an oxidizing biocide. Nonoxidizing biocides (such as calcium hypochlorite and sodium hypochlorite) interfere with normal organism metabolism. Both types are used to control microbiological activity in a wide range of commercial cleaning, environmental hygiene, disinfection, industrial, and water-treatment activities, often in combination.

# 4. CONCLUSION

Water treatment transforms raw surface water and groundwater into safe water for drinking, industry, medicine, and many other uses. Water treatment involves three types of processes: physical processes, such as settling and filtration; chemical processes, such as disinfection and coagulation; and biological processes, which may include, for example, the use of activated sludge (bacteria).

# 12

# DISTRIBUTION OF WATER

## 1. INTRODUCTION

After treatment, water usually is sent to finished water storage or directly into the distribution system or transmission lines that operate under a pressure head established by pumping (in relatively flat terrain) or by gravitation potential (which is possible in steep terrain). Water distribution systems are large networks of storage tanks, valves, pumps, and pipes that allow finished water to travel from the drinking-water treatment plant to the homes and businesses in your community. We'll look at the major types of water system components and different methods used for water distribution.

## 2. WATER SYSTEM COMPONENTS

Two basic types of water-supply systems manage the water pressure in most distribution networks. The goals are to reduce leaks, bursts, and interruptions to our water supply and to provide required water pressure to fire hydrants. Water supply components are important to the fire service, fire protection engineers, and city managers.

The two types of water systems that supply water under pressure are gravity feed systems and pumping pressure systems.

## ■ 2.1. GRAVITY FEED SYSTEMS

A gravity feed system is a basic water-movement system. One example is a rainwater harvesting system, where water is collected from the roof of a building (or as surface water runoff, in certain circumstances) and is pumped from the main holding tank (aboveground or belowground) to a header tank in the building's loft. There is usually a cold tank in the loft and a heated water tank on a lower level. The greater the drop (called the "head" of water) from the bottom of the cold-water tank to the top of the showerhead, the better the shower's pressure.

## ■ 2.2. PUMPING PRESSURE SYSTEMS

A pumping pressure system consists of a centrifugal pump (or several pumps), a pressure-accumulating tank, suction and discharge piping, electrical power, a pressure connector, regulatory and protective fittings, measurement and control devices (pressure and water gauge), and a compressed air installation equipped with a compressor and a driver (Kalenik 2009). The system must take water from a supply source, pass the water through a treatment plant, and then transport the water into the distribution system.

Community water systems are classified according to how water is sourced:

> ❱ High or low reservoirs that hold impotable water for gravity feed.

> ❱ Pumping station systems that use groundwater from streams, rivers, canals, man-made or natural lakes, and other special provisions for impound water. Pump stations stabilize and regulate flow all the way to the sewage treatment plant.

> ❱ Well water extracted with a pump. Based on the difference in elevation between the treatment facility and the community to be served, the water may flow by gravity through the distribution systems, or there may be the need for another pumping station.

> ❱ A combination of gravity flow and pump stations to transport the water from the source point to all of the water demand points on the distribution system.

# 3. METHODS OF WATER DISTRIBUTION

Water is dispersed throughout the distribution system according to local conditions or to regulations and requirements that influence water system design.

## ■ 3.1. GRAVITY DISTRIBUTION

Gravity distribution systems take advantage of natural elevation differences, distributing water from the service reservoirs to customers by gravity via a network of water mains. The pressure in the system is generally sufficient to provide a direct supply to six or seven stories above street level. It may be necessary to divide very tall buildings into distribution zones to keep the water pressure within proper limits. That is, upper floors are supplied from roof tanks filled by their own pumping systems. Since the static pressure from the water column leading all the way up to the roof might be excessive, the bottom zone may be upfed. Or, an intermediate tank halfway up may supply water to the lower half of the building, and a rooftop tank may supply the top half of the building (Bradshaw 2006).

## ■ 3.2. PUMPS

Many kinds of pumps are used in distribution systems for different applications. Low-lift pumps lift surface water (e.g., for the irrigation and drainage of land) and move it to a nearby treatment plant. They move large volumes of water at relatively low discharge pressures. Pumps that allow deep water pumping (e.g., potable water from wells) are called high-lift pumps. These operate under higher pressures. Pumps that are designed to smooth out water pressure in areas where the flows are highly variable are called booster pumps.

The pumps are usually centrifugal pumps (or rotodynamic pumps) or positive-displacement pumps. The centrifugal pump is the most widely used type in the world. In it, a rapidly rotating impeller adds energy to the water and raises the pressure inside the pump casing. The flow rate through a centrifugal pump depends on the pressure against which it operates. The higher the pressure, the lower the flow or discharge. Centrifugal pumps can be used as:

> ❭ Inline pumps

> ❭ End-suction pumps

❱ Double-suction pumps

❱ Vertical multistage pumps

❱ Horizontal multistage pumps

❱ Submersible pumps

❱ Self-priming pumps

❱ Axial flow pumps

❱ Regenerative pumps

Positive-displacement pumps operate by alternately filling a cavity and forcing a fixed volume of liquid from the inlet pressure section of the pump into its discharge zone. The water literally is pushed (displaced) from the pump casing. The flow capacity of a positive-displacement pump is unaffected by the pressure of the system in which it operates (Curley 2010). Positive-displacement pumps work as:

❱ Reciprocating pumps (with piston, plunger, and diaphragm)

❱ Power pumps

❱ Steam pumps

❱ Rotary pumps (gear, lobe, screw, vane, regenerative or peripheral, and progressive cavity).

## ■ 3.3. STORAGE TANKS

Distribution storage tanks, a familiar sight in many communities, serve two basic purposes: equalizing storage and emergency storage. Equalizing storage is the volume of water needed to satisfy peak hourly demands in the community (Curley 2010). Water can be pumped into the tank during periods of low demand and pumped back into the distribution system at peak demand. Water in a distribution storage tank may also be needed for fighting fires (storage tanks are economical and reliable), cleaning up accidental spills of hazardous materials, or other community emergencies, such as power outages, breaks in large water mains, problems at treatment plants, and so on. The capacity of a distribution storage tank is designed to be about equal to the average daily water demand of the community. In addition, storage tanks can provide the water pressure in the distribution system.

There are two types of storage tanks: ground water tanks and elevated water tanks.

**Ground water tanks** are either steel or concrete. The bottom of the water in them is at or near ground level. These tanks may receive water from a well or from surface water, allowing a large volume to be stored and used during peak demand cycles. The risk of freezing increases as more of a tank's surface is exposed to cold weather. Hodnett (1981) has addressed heating of tanks.

The water in a ground water tank is not put under a significant amount of pressure unless the tank is located at a higher altitude (on top of a hill, for example). Any pressure in a ground tank must be maintained through direct pumping, which can be costly.

**Elevated water tanks** are also known as water towers. They do not require the continuous operation of pumps, as gravity maintains the water pressure into the distribution system.

The locations of elevated water tanks can equalize water pressures in the distribution system, though precise water pressure can be difficult to manage. The pressure of the water flowing out of an elevated water tank depends on the depth of the water in it. A full elevated water tank may provide high pressure and a nearly empty elevated water tank may not provide enough pressure.

# 4. WATER DISTRIBUTION SYSTEMS

Water distribution systems typically have three types of pipeline systems for distributing water to demand points (consumers):

> **Primary feeders** are large pipes with diameters ranging from twelve inches in small cities to sixty inches in large cities. Primary feeders move water from the source, supply, or storage area to the secondary feeders (Couvillon 1993).

> **Secondary feeders** are smaller than primary feeders and transport water along the major streets of the community.

> **Distributor mains** are smaller than secondary feeders. They supply water directly to fire hydrants and homes. The minimum pipe size should be twelve inches in diameter on a principal street, with eight-inch mains cross-connected every six hundred feet in business districts and six-inch mains cross-connected in residential areas (Couvillon 1993).

# 5. CONCLUSION

A water distribution system is designed to deliver a sufficient quantity and quality of water to meet customer requirements. Typically, this is through water system components of a gravity feed system or a pumping pressure system. Water is delivered by way of pumps and motors, service pipes, storage tanks or reservoirs, and related equipment.

# Part IV:

## WATER USE

# WATER AND ITS EFFECTS ON LIFE

## 1. INTRODUCTION

Water makes up a majority of most organisms. It is a critical component in the processes of life. Most organisms are more than half water, while some organisms are 95 percent water. Let's look at the effects of water on life (biology) by understanding why water is so important to it.

## 2. WATER AND LIFE

Water and life are inseparable. No known living thing can function without water, and there is life wherever there is water on Earth (Rothschild and Mancinelli 2001). Civilizations and religions have recognized this link throughout history, and so do scientists today (Henry 2005). Of all the important conditions for terrestrial life, liquid water is the most important. No other liquid can replace it. Water is fundamental for enzyme action and for formation of the three-dimensional structure of proteins (Tuena de Gómez-Puyou and Gómez-Puyou 1998). Water is not only a solvent in which organic chemicals can be transported and react with each other; it is a vital ingredient in photosynthesis, the process that converts sunlight into sugars and thus provides the energy source for the majority of life.

$H_2O$ is the second-most common molecule in the universe (behind hydrogen, $H_2$), the most abundant solid material, which is fundamental to star formation. Nevertheless, water needs particularly precise conditions to exist as a liquid, as it does on Earth.

Without water, no existing life could have developed. How, exactly, life did develop is a question that has interested many people since well before the early experiments producing amino acids from simpler molecules by electric discharge in aqueous systems (Miller 1953). More recently, various theories have been propounded but without a consensus—except for the key involvement of liquid water (Trevors and Pollack 2005).

Water has particular properties that cannot be found in other materials and that are required for life-giving processes. The hydrogen-bonded environment that is particularly evident in liquid water brings about these properties (Fisenko and Malomuzh 2008).

Hydrogen bonds are roughly tetrahedrally arranged; a central atom is surrounded by four substituents, each at the corner of a tetrahedron (Figure 17). The local clustering can expand, decreasing the molecule's density. Low-density structuring naturally occurs at low and super-cooled temperatures, giving rise to many physical and chemical properties unique to liquid water. If aqueous hydrogen bonds were somewhat stronger than they are, water would behave similar to glass. If they were weaker, water would be a gas and could only exist as a liquid at sub-zero temperatures (Chaplin 2010).

From a biological standpoint, water is more than important; it's vital—both as a superb solvent (other substances regularly and easily dissolve into it, including body wastes) and as an essential part of many metabolic processes (chemical processes in the body). Without water, there would be no living organisms on the planet. The first sign that life is possible on other planets would be water. Where there is water, there is life, and where water is scarce, life struggles. Living cells are about 70 percent water, and water is needed for their continued functionality. Humans are basically an aqueous solution: water and organic compounds. To be deprived of water does not bode well.

**Metabolism** represents the sum of all reactions in a living system (organism), consisting of two simultaneous and constant processes: catabolism and anabolism. In other words, the total change in the system results from breaking down and building up. In **catabolism**, water is used as a solvent to break bonds and make smaller molecules, such as glucose and amino acids. Without water, these particular metabolic processes could not exist. In **anabolism**, water is removed from molecules (molecules are a stable group of two or more atoms) to make larger molecules, such as starches or proteins for storage.

Water is fundamental to photosynthesis and respiration in plants (Figures 35, 36, and 37).

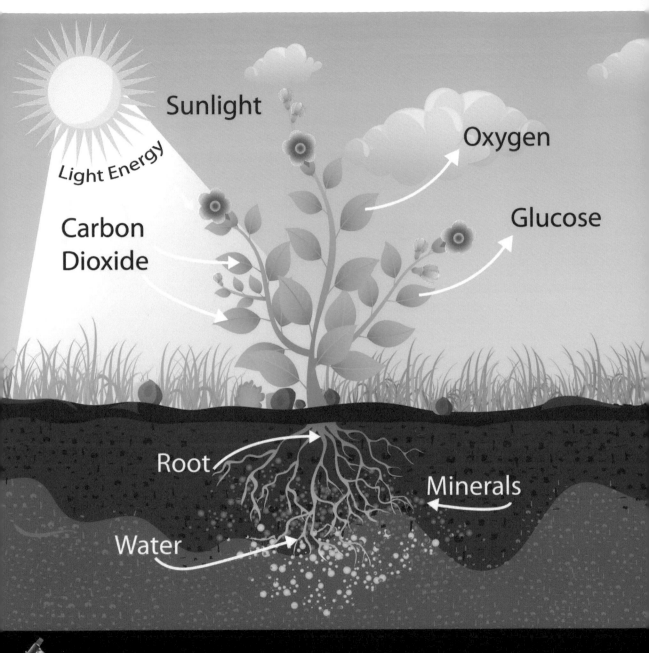

Figure 35. Photosynthesis in Plant Process

Figure 36.
Water is Used in
Photosynthesis:
Dambri Waterfall
in Central
Highland,
Vietnam

Figure 37. Water Photosynthesis/Respiration: Swans, Pitsford Reservoir, Northamptonshire, England, United Kingdom

**Photosynthetic** cells use the sun's energy to split off water's two hydrogen atoms from its oxygen atom. The hydrogen is combined with carbon dioxide ($CO_2$) that plants absorb from air or water to form glucose, which is a type of sugar that plants use as food. The plant releases the oxygen as "waste," and humans and animals breathe it. In cellular respiration, glucose is broken down to yield carbon dioxide and water, and the energy from this process is stored as adenosine triphosphate (ATP) molecules.

We've already noted that water is central to acid-base neutrality and enzyme function. An acid, which can be defined as a donor of a hydrogen ion ($H^+$—a proton), can be neutralized by a base (such as a hydroxide ion, $OH^-$), which is a proton acceptor, to form water.

$$H^+ + (OH)^- = HOH = H_2O = water$$

We already know that pure water is considered to be of neutral pH. (It is neither acid nor alkaline, defined as having a pH of 7.) Therefore, water is the reference point for acids and bases. Surface water systems typically range from pH 6.5 to 8.5, while groundwater systems range from pH 6 to 8.5. Water with a pH less than 6.5 is commonly called "soft." It is acidic and corrosive. It may contain elevated levels of toxic metals and can cause corrosion to piping. Water with a pH higher than 8.5 may taste bitter and soapy, though it may or may not be associated with health risks.

**Water turbidity**, or cloudiness, can affect fish and aquatic life. With excessive turbidity, water loses its ability to support a diversity of aquatic organisms. Water becomes warmer as suspended particles absorb heat from sunlight, and less light penetrates the water, resulting in even further drops in oxygen levels. These conditions cause reduced photosynthesis and decrease food availability, making it impossible for some forms of aquatic life to survive.

## 3. CONCLUSION

Water is life. Water molecules have unusual chemical and physical properties and unusual biological properties, which have many effects on life. It plays infinite roles in the lives of every organism, including humans, as we will explore further in the next chapter.

# 14

# WATER AND HUMAN CIVILIZATION

## 1. INTRODUCTION

Water is the most powerful substance on Earth. It is the deliverer of life. It cleanses and nourishes every living thing. Water has shaped human civilization since the beginning. Our earliest ancestors were hunters and gatherers. Then we created agriculture to produce enough food to stay in one place and form small cities. Civilizations grew up around water for agricultural reasons.

Over time, Egyptians, Mayans, and other ancient cultures built their economies around water—the spring flooding that produced the harvest for the coming year. Middle Eastern, Asian, and European civilizations all had similarly "hydrocentric" economies.

Libyans, Romans, and Greeks learned to move water through aqueducts and man-made tunnels to their metropolises using the power of gravity.

Fast-forward to modern times. We have learned a lot, such as how to control water and improve its quality, how the water cycle works, and how to extract groundwater. Water was, is, and will be the deliverer of life and human civilization. Without water, there would be no civilization at all.

Let's explore the human uses of water and the impacts of human civilization on water quality, and thus on human health.

# 2. HUMAN USES OF WATER

Water is our most precious resource. Water is vital to life. Humans are made up mostly of water. All humans will die if they don't have water.

■ **2.1. AGRICULTURE**

The main source of food for the population of the world is agriculture (Figure 38).

Figure 38. Rain-fed Agriculture: Olive Groves in Jaen, Andalusia (Spain)

The most important use of water in agriculture is irrigation. Up to 70 percent of the water we take from rivers and groundwater is used for irrigation. Rainfed agriculture is the set of farming practices that rely on rainfall for water, providing much of the food consumed by poor communities. Worldwide, rainfed agriculture is practiced in almost all hydroclimatic zones (Figure 38). In temperate regions with relatively reliable rainfall and productive soils, and in the subhumid and humid zones of tropical regions, rainfed agriculture can have some of the highest yields.

Since 1950, increases in yield have come from what is commonly called the "green revolution." Between 1950 and 1970, this approach resulted in dramatic increases in crop yields, mainly in more developed countries. During the 1970s, the construction of irrigation systems dramatically increased. However, the rate of growth began to decrease in both developed and developing countries in the 1980s. Irrigation has been a major factor in increasing crop productivity in many countries. It directly raises the productivity of land by providing sufficient water supply to raise the yield per hectare per crop. It also allows a second crop to be grown during the dry season when yields are potentially higher.

On the negative side, irrigation causes salinization of irrigated land, mostly in arid and semiarid regions. It increases the possibility that fertilizers and pesticides will infiltrate the groundwater or run off into nearby streams, causing groundwater contamination salinity. (Fertilizers contain nitrates, which are readily soluble in water and are very hard or impossible to remove from it.) In addition, the world population is predicted to grow from around 6 billion people today to 8.3 billion by 2030 (UN 2013). Agriculture needs to find a way to use less water or to use water more efficiently.

## ■ 2.2. DRINKING

Every human needs to drink a healthy amount of water (Figures 39 and 40).

Figure 39. Glass of Drinking Water

Figure 40.
Filling Glass
of Drinking
Water

### ■ 2.2.1. ROLE OF WATER IN HUMAN LIFE AND THE OPERATION OF THE HUMAN BODY

Water is the most essential substance for human life. The human body cannot survive without water because the body can't store it. The average body can survive only a week without water, depending on the weather. The adult human body is 60 percent water (Cook 2004). A human embryo is more than 80 percent water. A newborn baby is 74 percent water (Glenn 2013). The blood is more than 80 percent water; the brain, about 75 percent; saliva, 98 percent. The lungs are 90 percent water; the heart, 75 percent; perspiration, 98 percent. Bones are 22 percent water; and muscles, 75 percent. The human liver is an amazing 96 percent water, and the kidneys, 83 percent (Figure 41). It is clear that understanding water's important role in maintaining health is vital.

Water is the medium for various enzymatic and chemical reactions in the body. It moves nutrients, hormones, antibodies, and oxygen through the bloodstream and the lymphatic system. The proteins and enzymes of the body function more efficiently in solutions of lower viscosity (the internal resistance of fluids to shear forces); this is true of all the receptors in the cell membranes (Swilling 2004).

Water is the fundamental solvent for all biochemical processes in our bodies, and it regulates all functions, including the activity of everything it dissolves and circulates (Figure 42).

Water helps to regulate and to maintain body temperature by acting as a thermostat; one's health depends on keeping body temperature within a very narrow range. Perspiration helps to cool the body. It dissipates excess heat, especially in hot weather or during exercise, when increased sweating is more visible.

Water is essential to the proper chemical functioning of every single cell and metabolism in the human body. The digestion of protein and carbohydrates into usable and absorbable forms depends on water as part of the chemical reaction. Water serves as a solvent for nutrients and delivers them to cells.

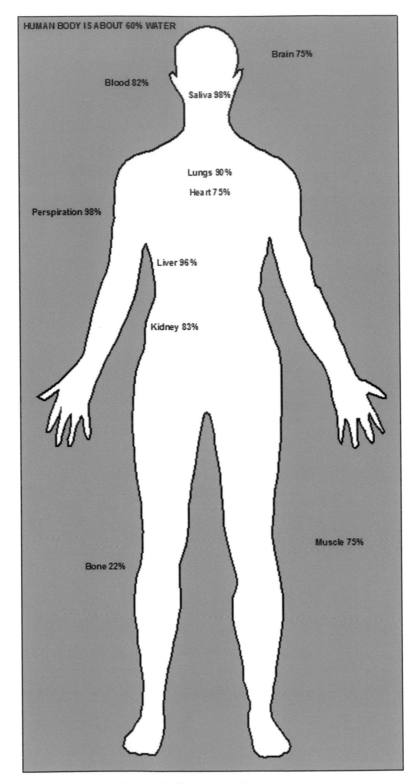

Figure 41. Water in the Human Body

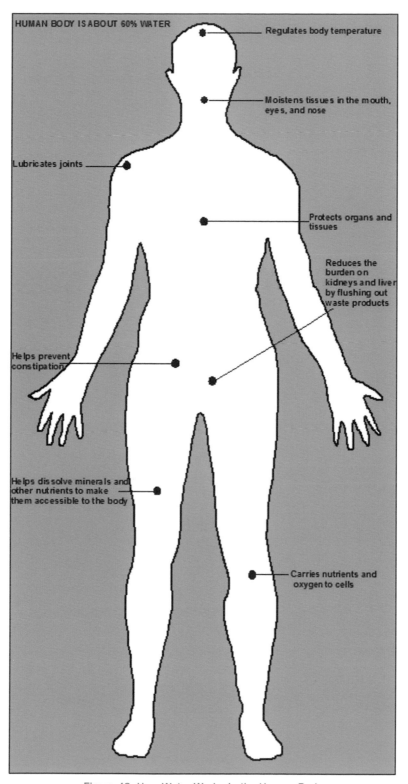

Figure 42. How Water Works in the Human Body

Water is essential for the efficient elimination of waste products (uric acid, urea, and lactic acid, all of which must be dissolved in water) through the kidneys. When there isn't sufficient water, those wastes are not effectively removed, which may result in damage to the kidneys. Water is essential in moving waste matter through the colon, as well. Drinking enough water (six to eight eight-ounce glasses per day) prevents and alleviates constipation.

Water is necessary for consuming and digesting food. It forms the base for saliva and serves to lubricate the alimentary canal (the digestive system) so that food and waste move easily through the body.

Water lubricates joints; water forms the fluids surrounding joints and bones, providing cushioning. Water plays an important role in the relief of common ailments, such as headaches and hypertension. Moreover, drinking enough water helps prevent the risk of bladder and colon cancer, and some studies have suggested that water may also decrease the risk of breast cancer.

## ■ 2.2.2. WHAT HAPPENS IF WE DON'T DRINK ENOUGH WATER?

Not drinking enough water can have negative effects on the body and the way it operates. Water consumption helps the body breathe because it moistens the lungs. Lung functions can use up to a pint of water every day, decreasing the body's moisture level through exhalation. When that liquid isn't replaced, breathing is more difficult.

Not drinking enough fluids can cause excess body fat, poor muscle tone, and a decreased ability to digest food. It may also encourage low blood pressure and a rapid heart rate. The body can become dehydrated, which can cause irritability, anger, and anxiety. People can become more easily stressed and experience other accompanying symptoms, such as headaches. In addition, a dehydrated body can have a dry and sticky mouth, sunken eyes, insufficient tears, little or no urine output, and lethargy.

A week without water results in death.

WATER AND HUMAN CIVILIZATION

## ■ 2.2.3. WHAT IS THE ROLE OF WATER IN DIET AND WEIGHT LOSS?

Initial weight loss is largely due to the loss of water, and you need to drink an adequate amount to prevent dehydration. Drinking an adequate amount of water will actually decrease your tendency to retain fluids.

If you eat a high amount of fiber, you need additional water. Fiber without adequate fluids can cause constipation instead of helping to eliminate it. Burning calories and fat creates waste products that require an adequate supply of water to flush out of your body through the kidneys.

Approximately 80 percent of our water intake comes from drinking water and other beverages, but many foods have high water contents. Your body may signal that it is hungry to get more water through food. But excess food supplies excess calories in addition to water.

Drinking water with a meal can help you feel full and therefore satisfied with eating less. However, drinking water alone may not have this effect. To feel satiated, the body needs calories and nutrients.

When you are working out, you're more likely to lose water through your breath and sweat, so be sure to drink enough. When you exercise, your muscle mass increases, allowing your body to burn more fat. Muscle is made up of more water than fat is, so water becomes even more important as you become more active; dehydration slows down the fat-burning activity of muscles. Water plays an important role in maintaining muscle tone and lubricating the joints, helping to reduce muscle fatigue and soreness during exercise.

Dehydration causes a reduction in blood volume, which causes a reduction in the supply of oxygen to the muscles. This can make you feel tired.

Caffeine (which can be found in coffee, tea, chocolate, and some sodas) and alcoholic beverages have a diuretic effect. For every cup of coffee you drink, you need an additional cup of plain water to counteract this effect.

## ■ 2.2.4. HOW MUCH WATER DO YOU NEED?

Everybody loses water daily through breath, perspiration, urine, and bowel movements. For the body to function properly, it needs at least enough water daily to replace the water it loses through these functions. (Some say this is at least eight eight-ounce glasses of fluid per day.) You need to replace even more lost water when it is very hot outside, during vigorous exercise, when you drink many caffeinated or alcoholic beverages, when you have a fever or lose water through vomiting or diarrhea.

### ■ 2.2.5. TIPS ON DRINKING ENOUGH WATER

Here are some tips to help you drink enough water every day.

> **Don't wait until you feel thirsty.** Thirst is your body's warning sign. Drink water before you feel thirsty, because by then, the body is already dehydrated. Drinking smaller quantities throughout the day is better than gulping down a large quantity at once.

> **Carry a water bottle with you**. When you are at work or occupied during the day, it is often easy to forget to drink regularly. So get a water bottle ready every morning and keep it with you.

> **Drink as much water as you like.** Your body is 60 percent water, so keep it in good working order. Perspiration, urination, and even breathing are all ways in which the body expels water permanently. So drink as much water as you want throughout the day. It's difficult to drink too much; your stomach has a natural limit.

> **Drink more water when it is hot.** When the outside temperature is 30°C or above, you should drink more, because the body dehydrates more rapidly due to perspiration.

> **Drink water during exercise.** Drink water regularly during and after exercise. When you increase your physical efforts, your body sweats more.

> **Drink enough when you are ill.** Your body needs more water when it is ill with fever, diarrhea, or vomiting because these sicknesses make your body lose large quantities of

water, possibly causing dehydration. For example, if you have a fever, drink a half liter more of water for every degree your temperature is higher than 37°C (98.6°F).

> **Eat more fruits that are rich in water.** Eat oranges, watermelon, pears, etc. This is a good way to take in more water.

> **Vary what you drink.** In your sixty-four ounces of water per day, you can mix it up a bit with tea, coffee, herbal infusions, broths, and light soup, all of which hydrate you.

> **Give your water a delicious flavor boost.** To make your water super yummy, you can infuse it with lemon, cucumber slices, ginger, or berries. Add a slice of lemon and leave it in your glass to give it a boost.

## ■ 2.3. WASHING

Water is used to form solutions and emulsions that are useful in various washing processes. Many industrial processes rely on reactions using chemicals dissolved in water, using the suspension of solids in water slurries, or using water to dissolve and extract substances. Of course, washing is important for personal hygiene (Figure 43).

## ■ 2.4. TRANSPORTATION

**Water transportation** is the process of moving people, goods, etc. by barge, boat, ship, or sailboat over a sea, ocean, lake, canal, river, etc. Water transportation also is the intentional movement of water over large distances. It is an important part of the world economy.

## ■ 2.5. CHEMICAL USES

In chemical reactions, water is the most abundant solvent and compound on Earth's surface. It is the only naturally occurring, inorganic compound that is a liquid at room temperature. It is an excellent solvent, capable of dissolving a greater variety of substances than any other liquid. The high polarity of water allows it to be attracted to many other different types of molecules, such as salt (NaCl), for example.

Figure 43.
Washing
Hands in
Water

138

Water is abundant and has many properties that make it a desirable solvent. It is:

> Polar, and therefore relatively easy to separate from other polar solvents

> Nonflammable and incombustible

> Cheap and widely available

> Odorless and colorless (making contamination easier to recognize)

> Conveniently separable from most organic substances due to its sufficiently different density

> Of high thermal conductivity, heat capacity, and heat evaporation, making exothermic reactions more easily controllable (Dave et al. 2005).

## ■ 2.6. HEAT EXCHANGE

If water is heated beyond its boiling point, it vaporizes into steam (water in the gaseous state), but not all steams are the same. Their properties vary greatly depending on the pressure and temperature to which they are subject. Condensing steam is a particularly efficient heating fluid because of the large heat of vaporization. A disadvantage is that water and steam are somewhat corrosive.

In almost all electric power stations, water vaporizes and drives steam turbines to drive generators. It is also the coolant.

## ■ 2.7. FIRE EXTINCTION

Water is the most efficient substance for extinguishing most fires. The fire fighter must understand its advantages and disadvantages in particular circumstances. It is dangerous to use water on fires involving oils and organic solvents because many organic materials float on water, and the water can spread the burning liquid.

In addition, a steam explosion is a potential hazard if water is superheated (it can be rapidly heated by fine, hot debris within it in confined spaces) or heated by the interaction of molten metals. The reaction between steam and hot zirconium can induce a hydrogen explosion. Pressure vessels that operate above atmospheric pressure can also decompose the water and produce conditions for a steam explosion. Such explosions occurred in the Chernobyl disaster when the extreme overheating of the reactor core caused water to flash into steam, although the water involved was from the reactor's own water cooling system (not from efforts to extinguish the fire).

## ■ 2.8. RECREATION

Humans use water for many different forms of recreation, including exercise and sports such as swimming, boating, surfing, white water rafting, ice hockey, and ice-skating (Figure 44). Other activities, such as picnics at the beach, hiking, nature viewing, and hunting may not involve contact with water, but they may still require water or be enhanced by proximity to it. Some like to go fishing or engage in snow sports like skiing, sledding, snowmobiling, or snowboarding.

## ■ 2.9. WATER IN INDUSTRY

Industry depends greatly on water, just as agriculture and homes do. The water industry provides drinking water and wastewater services (including sewage treatment) to households and industry.

The volume of annual water use by industry is increasing at an enormous rate—from 752 $km^3$ per year in 1995 to a likely 1,170 $km^3$ per year by 2025 (Schroeder 2004).

Freshwater is needed for industries to survive, but it is also needed for humans to survive. Every government in the world has programs to distribute water to the needy at no charge. In some cities, such as Hong Kong, seawater is used extensively for flushing toilets to conserve freshwater resources.

## ■ 2.10. INDUSTRIAL APPLICATIONS

Water is used in power generation. Hydroelectricity is electricity obtained from hydroelectric power, which must be one of the oldest methods of producing power. Water falling by the force of gravity can be used to turn turbines and generators that produce electricity. Typically, a dam is constructed on a river to create hydroelectric power. Water builds up behind barriers, creating artificial lakes known as reservoirs. Hydroelectricity is cheap and nonpolluting. Many small hydroelectric facilities are considered renewable energy sources.

Figure 44. Water Recreation, Feluka in Navigation on The River Nile, Luxor, Egypt

Pressurized water is used in water blasting and water-jet cutting, a technology that directs high-pressure streams of water for precise cutting. Plain water under pressure makes this technology relatively safe, and it is not harmful to the environment.

Water also is used to cool machinery and saw blades to prevent overheating.

Water is also an important raw material because it is needed in many industrial processes and machines, such as the steam turbine and heat exchanger, in addition to its use as a chemical solvent. Discharge of untreated water from industrial uses is pollution. Pollution includes discharged solutes (chemical pollution) and discharged coolant water (thermal pollution). Many processes in the chemical and petrochemical industry need pure water and require a variety of purification techniques in both water supply and discharge.

## ■ 2.11. FOOD PROCESSING

When we talk about food, we automatically talk about water, because without water, no food can exist. Water plays many critical roles within the field of food science. A food scientist must understand them to ensure the success of food products.

Solutes such as salts and sugars found in water affect its physical properties. Solutes increase the boiling point and decrease the freezing point. For example, one mole of sucrose (sugar) per kilogram of water raises the boiling point by 0.51°C, and one mole of salt per kilogram raises it by 1.02°C (Vaclavik and Christian 2007). Solutes in water also affect water activity, which affects many chemical reactions and the growth of microbes in food (DeMan 1999). Water activity is defined as the ratio of the vapor pressure of water in a material to the vapor pressure of pure water at the same temperature. Solutes in water lower water activity. This is important, because most bacterial growth ceases at low levels of water activity (DeMan 1999). Not only does microbial growth affect the safety of food but also its preservation and shelf life.

Water hardness is an aesthetic quality of water, caused mostly by the minerals calcium (Ca) and magnesium (Mg). However, its classification is based on the level of concentration of calcium carbonate ($CaCO_3$). We generally define soft water as having less than 75 mg/L calcium carbonate and hard water as having greater than 150 mg/L calcium carbonate (Sawyer and McCarty

1967). The hardness of water may be altered or treated by using a chemical ion exchange system. Water hardness has a major effect on pH and pH stability, which plays a critical role in food processing. For example, hard water prevents successful production of clear beverages. Water hardness also affects sanitation; with increasing hardness, water loses effectiveness as a sanitizer (Vaclavik and Christian 2007).

# 3. CONCLUSION

From the smallest bacteria to the largest trees, all forms of life need water to survive. We could not survive for more than a few days without drinking water or getting water from the foods we eat. We also need water for agriculture, industry, household uses, and recreation. Water is continually cycled and recycled through the environment. Civilizations are constrained directly by the quality and quantity of available safe drinking and subsistence water. They are also constrained indirectly by the influence of water on food, energy, transportation, and industry.

# MANAGING WATER SCARCITY

## 1. INTRODUCTION

Water is essential for all socioeconomic development and for maintaining healthy ecosystems. As a growing world with more people and more activities puts tremendous strain on Earth's waters, the cost and effort to build or even maintain access to water will increase. Therefore, the problem of water scarcity is growing, and need for improved water management.

Here, we'll look at water pollution and its effects on human health, discuss the risks of water scarcity for human life, and give an overview of managing water scarcity through water policy and government and organization actions to prevent and mitigate water scarcity with the goal of moving toward a water-efficient and water-saving economy.

## 2. WATER POLLUTION

Water pollution is the contamination of water bodies, very often by human activities. Water pollution occurs when pollutants are discharged (directly or indirectly) into water bodies without adequate treatment to remove harmful compounds.

Pollutants in water include a wide spectrum of chemicals (including toxic ones) and pathogens, and bring about physical, chemical, or sensory changes. Many classes of pathogens excreted in feces can initiate waterborne infections, including bacterial pathogens (enteric and aquatic), enteric viruses, and enteric protozoa, which are strongly resistant in the water environment and to most disinfectants (Leclerc et al. 2002). Alteration of water's physical chemistry includes changes in acidity (change in pH), electrical conductivity, temperature, and eutrophication.[24] Eutrophication can be natural but can also result from human activity (cultural eutrophication from fertilizer runoff and sewage discharge) and is particularly evident in slow-moving rivers and shallow lakes. Increased sediment deposition can eventually raise the level of the lake or riverbed, allowing land plants to colonize the edges, eventually converting the area to dry land (Lawrence et al. 1998). Salts, oil, gasoline, and chemicals, including fertilizers and pesticides, can contaminate groundwater and make it unsafe and for human use.

Interactions between surface water and groundwater are complex. Consequently, groundwater aquifers are susceptible to contamination from either surface water or sources that may not directly affect surface-water bodies. The distinction of point versus nonpoint source may be irrelevant. A spill of a chemical contaminant on soil, located away from a surface-water body, may not necessarily create point-source or nonpoint-source pollution but may contaminate the aquifer below (a toxin plume). Analysis of groundwater contamination may focus on soil characteristics and hydrology as well as on the nature of the contaminants.

**Water Pollution and Disease.** Water pollution affects drinking water, rivers, lakes, and oceans all over the world, which may cause many diseases that are referred to as water pollution diseases. These could have serious health impacts. Freshwater contamination can result in health problems for everyone from pregnant women and newborns to the elderly and cancer patients undergoing chemotherapy.

The most common water pollution diseases involve poisoning episodes affecting the digestive system and human infectious diseases, including:

> Typhoid, usually transmitted by water or food in much the same way as cholera.

> Intestinal parasites that populate the gastrointestinal tract in humans and animals.

> Enteric and diarrheal diseases caused by bacteria, parasites, and viruses.

---

24 Eutrophication is the enrichment of bodies of freshwater by inorganic plant nutrients, such as nitrate and phosphate.

- Giardiasis, an infection of the digestive system caused by tiny parasites called *Giardia intestinalis*. Diarrhea is the most common symptom of giardiasis.

- Amoebiasis, caused by the protozoan *Entamoeba histolytica* (Stanley 2003). Amoebiasis is often asymptomatic but may cause dysentery and invasive extraintestinal disease.

- Ascariasis, caused by the parasitic roundworm *Ascaris lumbricoides*.

- Hookworm, a parasitic nematode worm that lives in the small intestine of its host. Two species of hookworms commonly infect humans: *Ancylostoma duodenale* and *Necator americanus*.

- Gastroenteritis, an infection of the stomach and bowel (large intestine).

- Stomach cramps and aches.

- Hepatitis, which is inflammation of the liver.

- Encephalitis, an uncommon but serious inflammation of the brain.

- Diarrhea.

- Respiratory infections.

- Vomiting.

- Endocrine (hormonal system) damage, including interrupted sexual development, thyroid system disorders, inability to breed, degraded immune function, mental impairment, decreased fertility, and increases in some types of cancers.

- Damage or injury to kidneys.

- Malaria, the world's most important parasitic infectious disease, is transmitted by mosquitoes that breed in fresh (or occasionally brackish) water.

❱ Liver damage and even cancer (due to DNA damage).

❱ Damage to the nervous, kidney, blood forming, heart, and immune systems.

❱ Less serious health effects, such as rashes, earaches, and pinkeye.

# 3. WATER SCARCITY

Water (more precisely, freshwater), is as important to us as air. However, less than 1 percent of the planet's water is readily available for drinking or for most agriculture. Freshwater is becoming scarce, and that is becoming one of the world's biggest problems. The lack of freshwater is currently one of the main obstacles to the economic development of many countries, a problem that is widespread around the globe. According to the United Nations, more than a billion people lack access to safe drinking water, and 2.6 billion are without basic sanitation (Bouzane 2010).

Water scarcity is not only the lack of enough water (quantity) but also the lack of access to safe water (quality).

Some of us cannot take potable drinking water for granted. In the developing world, finding a reliable source of safe water is often time-consuming and expensive. This is known as economic scarcity. However, in other areas, the lack of water is a more profound problem. There simply isn't enough. This is physical scarcity.

Hydrologists typically assess scarcity by looking at the population-water equation. An area is experiencing water stress when annual water supplies drop below 1,700 $m^3$ per person (IPCC 2007). When annual water supplies drop below 1,000 $m^3$ per person, the population faces water scarcity, and below five hundred cubic meters, it experiences "absolute scarcity" (UN-Water 2007).

Water scarcity is expected to increase because the distribution of precipitation in space and time is very uneven, leading to tremendous temporal variability in water resources worldwide (Oki and Kanae 2006). Second, the rate of evaporation also varies a great deal, depending on temperature and relative humidity, and this affects the amount of water available to replenish groundwater supplies. The combination of more intense rainfalls of shorter duration and increased evapotranspiration and irrigation is expected to lead to groundwater depletion (Konikow and Kendy 2005).

Pollution has intensified the water shortage problem. Acid rain is one major factor. In many places in the world, rivers, lakes, and groundwater are heavily polluted. Thus, some water-supply systems are too polluted to be suitable for human contact.

### ■ 3.1. QUANTITY OF WATER

Many countries in the world already struggle under existing water stress from pressures such as irrigation demands, industrial pollution, and waterborne sewerage. Sub-Saharan Africa has the largest number of water-stressed countries of any region. These pressures will be exacerbated significantly by climate change, which for many regions will result in increasing temperatures and reduced rainfall, further reducing the availability of water for drinking, household use, agriculture, and industry. Consequently, with the existing climate change scenario, almost half the world's population will be living in areas of high water stress by 2030, including between 75 million and 250 million people in Africa. In addition, water scarcity in some arid and semiarid places will displace between 24 million and 700 million people (UN-Water 2007).

According to the Comprehensive Assessment of Water Management in Agriculture, one in three people are already facing water shortages (2007). Over 1.2 billion are living in areas of physical water scarcity. (There is not enough water to meet all demands, and water is not distributed evenly.) And almost 1.6 billion face economic water shortage. (There is a lack of investment in water programs and water capacity.) These are extreme numbers. As our population continues to grow, there will be more problems. We face drastic measures to make sure everyone has access to water (DeCapua 2013).

The UN World Water Development Report from the World Water Assessment Program predicts that in the next twenty years, the quantity of water available to everyone will decrease by 30 percent (UN WWDR 2003). By 2025, 1.8 billion people will be living in countries or regions with absolute water scarcity, and two-thirds of the world's population could be living under water-stressed conditions (UN-Water 2007).

### ■ 3.2. QUALITY OF WATER

The quality of existing water supplies will become a further concern in some regions. Most of the geochemical and biochemical substances in water are acquired during its travel from the clouds to the rivers through the biosphere, the soils, and the geological layers. Changes in the amounts or patterns of precipitation will change the routes and residence times of water in the watershed, thereby affecting its quality. As a result, regardless of quantity, water could become unsuitable as a resource if newly acquired qualities make it unfit for the required uses.

About 1.2 billion people lack clean water (Donya 2007). The UN World Water Development Report from the World Water Assessment Program indicates that 40 percent of the world's inhabitants currently have insufficient freshwater for minimal hygiene. More than 2.2 million people died in 2000 from waterborne diseases (those related to the consumption of contaminated water) or drought. Waterborne infections account for 80 percent of the world's infectious diseases (Donya 2007).

About 2.4 billion people—half of the developing world—in rural and urban areas lack access to adequate sanitation services. Within twenty years, it is expected that an additional 2 billion will live in towns and cities, mainly in developing countries, and they will need sanitation. In developing countries, more than 90 percent of sewage is discharged untreated, polluting rivers, lakes, and coastal areas (Günter and Elke 2005). As direct consequences:

An estimated 1.8 million people die every year from diarrheal diseases (including cholera) attributable to lack of access to safe drinking water and basic sanitation. Of these, 90 percent are children under five, most of whom live in developing countries (World Health Organization 2004).

Schistosomiasis is estimated to affect more than 200 million people especially in rural and agricultural areas, leading to the loss of 1.53 million disability-adjusted life years, according to the World Health Organization (Gryseels et al. 2006).

Intestinal helminths (ascariasis, trichuriasis, and hookworm infection) plague the developing world due to inadequate drinking water, sanitation, and hygiene. An estimated 133 million people suffer from high-intensity intestinal helminth infections, and there are about 1.5 million cases of clinical hepatitis every year (World Health Organization 2004).

Malaria kills twice as many people every year as formerly believed, taking the lives of 1.2 million babies, older children, and adults (Boseley 2012).

# 4. MANAGING WATER SCARCITY

Water scarcity can largely be avoided with better water-management practices. The validity of traditional knowledge on water management and the practices derived from it have been studied and documented since the 1980s.

## ■ 4.1. PREVENTING WATER POLLUTION

The human body needs clean water for healthy living, yet millions of people worldwide are deprived of it. We can do many things to keep water cleaner.

*One person alone cannot save the planet's biodiversity, but each individual's effort to encourage nature's wealth must not be underestimated.*

—United Nations Environment Programme (UNEP)

You can prevent pollution of nearby rivers and lakes as well as groundwater and drinking water by following some simple guidelines in your everyday life.

> Dispose of trash in the correct waste bin. If there is none around, please take trash home and put it in your trash can. You can clean up any litter you see on beaches, at the riverside, and on and near water bodies.

> Never throw chemicals, oils, paints, or medicines down the sink drain or the toilet. Check with your local authorities to see if there is a chemical disposal plan for local residents.

> If you use fertilizers and pesticides for your gardens or farm, take great care not to overuse them. This will reduce runoff. Consider composting and using organic manure instead.

> Use environmentally responsible household products for laundry, household cleaning, and toiletries.

> If you live close to a water body, grow more plants in your yard so that rain drains fewer chemicals from your property into the water.

Many governments have strict policies and laws for industry, hospitals, schools, and commercial concerns to reduce the risk of water pollution. These include:

> Monitoring of water and vector-borne diseases, particularly in the aftermath of disasters;

> Capacity planning in health care and integration of health care and disaster risk management planning;

> Preventive health maintenance programs promoting healthy lifestyles and improved nutrition and hygiene.

## ■ 4.2. WATER HARVESTING

Water harvesting is the trapping of rainwater for reuse, usually before it reaches the aquifer. Humanity has harvested water almost as long as it has practiced agriculture. There is evidence that ancient Greeks were the first to harvest water. Some countries do not have adequate rainfall to use this technique (Manzoor 2011).

Rainwater harvesting is defined as a method for inducing, collecting, storing, and conserving local surface runoff for agriculture in arid and semiarid regions (Boers and Ben-Asher 1982). It provides drinking water, water for livestock, and so on during regional water restrictions (i.e., droughts) and allows groundwater levels to replenish.

Groundwater harvesting is the extraction of groundwater. Groundwater recharge areas should be identified and monitored. Communities should manage them to prevent external disturbances.

Floodwater harvesting is collecting and storing rainfall runoff that enters infiltration systems, watercourses, or sewers during or shortly after a rainfall event for irrigation use.

Many people in the world continue to rely on water harvesting practices. Some in our own society are discovering ways that rainwater harvesting can be used to relieve stress on the environment, save money, and recharge groundwater tables. Other groups have returned to water harvesting to relieve pressure on overburdened underground water tables or municipal water systems.

In the early 1990s, several studies on traditional water-harvesting infrastructures were published (Prinz and Wolfer 1999). These techniques of water harvesting survive today, particularly in arid and semiarid areas, such as North Africa and sub-Saharan Africa.

## ■ 4.3. SEAWATER DESALINIZATION

According to the United Nations Environment Programme (UNEP), only 0.3 percent of the world's total amount of water is clean enough to drink. As natural freshwater resources are limited and humans need huge amounts of supply water (drinking water) every day, seawater can be an important source as long as it is desalinated (has its salt and other minerals removed).

Distillation is the method that most completely reduces the widest range of drinking water contaminants. It is the only process that replicates the hydrological cycle. Salt water is heated, producing water vapor. Then the water vapor is condensed into freshwater. It is a simple evaporation-condensation-precipitation system.

Electrodialysis is an electrochemical separation process that uses electrical currents to force salt ions to move selectively through a membrane, leaving freshwater behind.

Reverse osmosis uses a semipermeable membrane to separate water and salts both in industrial processes and in producing potable water.

Desalinization may be a viable option to overcome water shortages in many countries, such as Saudi Arabia, the United Arab Emirates, and Qatar, whose programs have been quite successful. Most countries of the Middle East, including Syria, have adequate energy to power Desalinization plants, which make them a vital medium-term solution (Morvan et al. 2005).

Desalinization is still a costly water-supply option compared to natural water resources (e.g. drawing from groundwater or surface water) and is outside the reach of poor nations.

■ **4.4. WATER RESOURCE POLICY**

Institutional and governmental **water resource policies** promote awareness of water scarcity and water use. Recycling of wastewater (reusing it after removing contaminants from domestic, commercial, and institutional applications) could be a viable solution to local water shortages in many low-income countries.

Many organizations now promote water conservation. They raise awareness and conduct research into water resources and effective water policies. Such organizations include the American Water Resources Association, Water Aid, the International Water Association (IWA), and Water 1st International. The International Water Management Institute undertakes projects with the aim of using effective water management to reduce poverty. Water-related conventions include the International Convention for the Prevention of Pollution from Ships, the United Nations Convention on the Law of the Sea, the United Nations Convention to Combat Desertification (UNCCD), and the Ramsar Convention. World Day for Water takes place on March 22, and World Ocean Day on June 8.

The proportion of people in developing countries with access to safe water has improved from 30 percent in 1970 to 71 percent in 1990, 79 percent in 2000, and 84 percent in 2004. This trend is projected to continue (Lomborg 2001).

A 2006 United Nations report states that "there is enough water for everyone" but mismanagement and corruption hampers access to it (UNE and UNESCO 2006). In addition, donors in the water sector have not embraced global initiatives to improve the efficiency of aid delivery, such as the Paris Declaration on Aid Effectiveness, as effectively as have donors in the education and health sectors. As a result, multiple donors may be working on overlapping projects, and recipient governments may not be empowered to act (Welle et al. 2009).

Water scarcity is still growing despite the policies and actions of different societies and organizations. Some options for good water resource management include:

> Improved water-management systems, demand-side management, water conservation, and efficiency improvements in water utilization.

> Sustainable management of groundwater sources; protection of water quality; water reuse; climate-proofing of water supplies against salinity, floods, and storm surges.

> Development of redundant/emergency water supplies, integrated water-resource management, and disaster risk management.

## 5. CONCLUSION

Freshwater resources are an essential component of Earth's hydrosphere and an indispensable part of all terrestrial ecosystems. Water scarcity (quantity and quality) is one of the greatest problems faced by the 21st century world. Lack of access to clean water forces people to obtain drinking water from unsafe sources. Polluted drinking water causes health problems and leads to waterborne diseases. Managing water scarcity requires: preventing water pollution, increasing the use of water harvesting and seawater desalinization, and building better management policies in water-scarce (promote strong awareness, remove contaminants, and make many organizations).

# CONCLUSION

Water is among the principal components of the universe and is an element essential for life. Of the water on Earth, 97.5 percent is salt water. Less than 1 percent of all the water on Earth is freshwater, but it is the only part we can consume or use for economic development.

Water is the only substance on the planet that naturally occurs in three different states: solid, liquid, and gaseous, depending on the atmospheric pressure. Virtually all forms of life depend on water. Water helps to define the habitable zone for humans, plants, and animals. Water is perfect. It has many unique chemical and physical properties that are crucial to the existence of life. It is a great solvent.

The water cycle is how Earth's water recycles itself. The cycle includes precipitation, evaporation, transpiration, and condensation. Earth's water keeps changing from liquid water to vapor, and then back again, a cycle perpetuated by the sun's heat and gravity. The water cycle is the driving force behind weathering, erosion, and deposition.

Water supply engineering is concerned with the development of water sources and water quality, treatment, and distribution. The term is used most frequently with regard to municipal water works, but it also applies to water systems for industry, irrigation, and other purposes.

Life could not exist without water; it needs water's special properties. Without water, there would be no human civilization. We use water to drink, travel, fish, wash, cool down, cook, water plants, and more.

The world's water problems are neither homogeneous nor consistent; they vary significantly from one region to another even within a single country, from one season to another, and from one year to another. Water scarcity urges everyone to be part of efforts to conserve and protect the resource. Managing water scarcity calls for increasing the use of desalinization and purification technologies, building better management policies in water-scarce and increasing public awareness regarding water use and conservation. Many different kinds of organizations—from national governments to local community groups—play roles in water policy decisions. These initiatives are essential to ensure that the world's water can sustain future generations.

# SUMMARY

Water exists in many places in the universe and may be the principal element for life on Earth's surface. Its presence is a hallmark of the habitable zones for living things (humans, animals, and plants).

About 70 percent of Earth's surface is covered with water, but only about 1 percent of the water on Earth's surface is usable by living things (groundwater and surface water).

Water has several properties that make it unique among compounds and make it possible for all forms of known life to function. Water is a uniquely wonderful substance on Earth that naturally occurs in three different states: solid, liquid, and gaseous. All living things depend on water. When water resources are polluted, all forms of life are threatened.

Water scarcity is a function of cultural activities and human civilization. It is very important to highlight water's pivotal role as the essential ingredient for life and to develop a comprehensive framework for creating water-resource policies to manage water resources.

This book includes photographs of water and life to demonstrate to the reader water's properties and water's effects on life.

# BIBLIOGRAPHY

Aguilar, D. "Astronomers Find Super-Earth Using Amateur, Off-the-Shelf Technology." Harvard-Smithsonian Center for Astrophysics, 2009. http://www.cfa.harvard.edu/news/2009-24

Bagchi, B. "Water Dynamics in the Hydration Layer around Proteins and Micelles." *Chemical Reviews* 105, no. 9 (September 2005): 3197–3219.

Baumgartner, A. and E. Reichel. *The World Water Balance: Mean Annual Global, Continental and Maritime Precipitation, Evaporation and Run-Off.* New York: Elsevier Scientific Publishing Co., 1975.

Beven, K. "Robert E. Horton's Perceptual Model of Infiltration Processes." Hydrological Processes 18 (2004): 3447–3460. doi: 10.1002/hyp.5740.

Boers, T. M. and J. Ben-Asher. "A Review of Rainwater Harvesting." *Agricultural Water Management* 5 (1982): 145–158.

Boseley, S. "Malaria Kills Twice as Many as Previously Thought." 03/02/2012. http://www.theguardian.com/society/2012/feb/03/malaria-deaths-research

Boswinkel, J. A. "Information Note, 2000." International Groundwater Resources Assessment Centre (IGRAC), Netherlands Institute of Applied Geosciences, Netherlands, 2000.

Bouzane, B. "UN Declares Water, Sanitation as Human Rights." *Postmedia News*, July 28, 2010.

Bradshaw, V. *The Building Environment: Active and Passive Control Systems.* 3rd ed. Hoboken, N.J.: John Wiley and Sons. 2006.

Brown, M. E. "Pluto and Charon: Formation, Seasons, Composition." *Annual Review of Earth and Planetary Sciences* 30, no. 1 (May 2002): 307–345.

Cain, F. "Is There Water on Venus?" *Universe Today.* July 29, 2009. http://www.universetoday.com/36291/is-there-water-on-venus/

Cain, F. "Atmosphere of Mercury." *Universe Today.* March 12, 2012. http://www.universetoday.com/22088/atmosphere-of-mercury/

Chang, K. "Saturn Moon Has Water." *New York Times*, March 10, 2006, A7.

Chaplin, M. "Do We Underestimate the Importance of Water in Cell Biology?" *Nature Reviews Molecular Cell Biology* 7, no. 11 (November 2006): 861–866.

Chaplin, M. F. "Water's Hydrogen Bond Strength," In Water of Life: The Unique Properties of H2O, ed. Ruth M. Lynden-Bell, Simon Conway Morris, John D. Barrow, John L. Finney, and Charles L. Harper, Jr. CRC Press, Boca Raton (2010) pp. 69-86.

Cook, A. "Thought for the Week: Northwest & Merseyside edition." *Daily Post* (Liverpool, UK), July 17, 2004, p. 6.

Cosgrove, W. J. and F. R. Rijsberman. *World Water Vision: Making Water Everybody's Business.* London: Earthscan Publications, 2000.

Cottini, V., N. I. Ignatiev, G. Piccioni, P. Drossart, D. Grassi, and W. J. Markiewicz. "Water Vapor near the Cloud Tops of Venus from Venus Express/VIRTIS Dayside Data." *Icarus* 217, no. 2 (June 2011): 561.

Couvillon, A. R. *Fire Captain Written Exam Study Guide.* Hermosa Beach, CA: Information Guides, 1993.

Curley, R. *New Thinking about Pollution (21st Century Science).* New York: The Rosen Publishing Group, 2010.

Dave J. A., P. J. Dyson, and S. J. Tavener, *Chemistry in Alternative Reaction Media.* Hoboken, N.J.: John Wiley and Sons, 2005.

Davis, T. A. "Electrodialysis." *Handbook of Industrial Membrane Technology,* edited by M.C. Porter, chap. 8: 482–510. Park Ridge, N.J.: Noyes Publications, 1990.

DeCapua, J. "Will There Be Enough Water for Everyone?" *Voice of America News*, March 2013.

DeMan, J. M. *Principles of Food Chemistry.* 3rd ed. Gaithersburg, MD: Aspen Publishers, 1999.

Dennis, L. M., C. C. Julie, L. Jonathan, and V. J. Torrence. "Enceladus' Plume: Compositional Evidence for a Hot Interior." *Icarus* 187, no. 2 (April 2007): 569–573.

Diersing, N. "Water Quality: Frequently Asked Questions." Florida Brooks National Marine Sanctuary, Key West, FL. 2009.

Donya, C. A. "Pollution Causes 40 Percent of All Deaths." *The Nation's Health* 37, no. 8 (October 2007): 11.

Dornheim, M. A. "Liquid Water at Saturn." *Aviation Week & Space Technology* 164, no. 11 (March 13, 2006): 27.

Earle, S. A. "The Great Dying-out of The Big Fish," May 25, 2003. *The Los Angeles Times.* http://articles.philly.com/2003-05-25/news/25460327_1_big-fish-wild-fish-marine-life

Eddy, M. "Water on Mars." *Tulsa World*, January 24, 2004, A8.

Ehlers, E. and T. Krafft. *Understanding the Earth System: Compartments Processes, and Interactions.* New York: Springer, 2001.

Fewtrell, L. "Drinking-Water Nitrate, Methemoglobinemia, and Global Burden of Disease: A Discussion." *Environmental Health Perspectives* 112, no. 14 (October 2004: 1371–1374.

Fine, R. A., and F. J. Millero, "Compressibility of Water as a Function of Temperature and Pressure." *Journal of Chemical Physics* 59, (1973): 5529–5536.

Fisenko, A. I. and N. P. Malomuzh. "The Role of the H-bond Network in the Creation of the Life-giving Properties of Water," *Journal of Chemical Physics* 345 (2008): 164–172.

Frauenfelder, H., G. Chen, J. Berendzen, P. W. Fenimore, H. Jansson, B. H. McMahon, I. R. Stroe, J. Swenson, and R. D. Young. "A Unified Model of Protein Dynamics." *Proceedings of the National Academy of Sciences of the United States of America* 106, no. 13 (March 31, 2009): 5129–5134.

Fyall, J. "Drop in the Ocean? Scientists Link Comet Ice to Earth's Water." *The Scotsman*, June 10, 2011, 23.

Gehl, L. "Ask Dr. Cy Borg." *Odyssey*, 22, no. 3 (March 2013): 46.

Gleick, P. H., ed. *Water in Crisis: A Guide to the World's Freshwater Resources.* New York: Oxford University Press, 1993.

Glenn, E. "Water does much more than quench thirst." Philadelphia Tribune, 07/23/2013, Volume 129, Issue 71, p. 7B.

Goudie, A. S. and H. A. Viles. "The Nature and Pattern of Debris Liberated by Salt Weathering: a Laboratory Study." *Earth Surface Processes and Landforms* 9 (1995): 95–98.

Griffith, C. A., T. Owen, T. R. Geballe, J. Rayner, P. Rannou. "Evidence for the Exposure of Water Ice on Titan's Surface." *Science* 300, no. 5619 (April 2003): 628–630.

Groombridge, B. and M. Jenkins. *Freshwater Biodiversity: A Preliminary Global Assessment.* Cambridge, UK: World Conservation Press, 1998.

Gryseels, B., K. Polman, J. Clerinx, and L. Kestens. "Human Schistosomiasis." *The Lancet* 368, no. 9541 (2006):1106–1118.

Günter, L. and M. Elke. "Ecological Sanitation—a Way to Solve Global Sanitation Problems?" *Environment International* 31, no. 3 (April 2005): 433–444.

Hanor, J. S. "Hydrosphere." in Moore C. (ed.) McGraw-Hill Encyclopedia of Science and Technology, 9th edition, 2002.

Hansen, C. J., L. Esposito, A. I. F. Stewart, J. Colwell, A. Hendrix, W. Pryor, D. Shemansky, and R. West. "Enceladus' Water Vapor Plume." *Science* 311, no. 5766 (March 2006): 1422–1425.

Held, I. M. and B. J. Soden. "Water Vapor Feedback and Global Warming." *Annual Review of Energy and the Environment* 25, no. 1 (November 2000): 441–475.

Henniker, J. C. "The Depth of the Surface Zone of a Liquid." *Reviews of Modern Physics* 21, no. 2 (1949): 322–341.

Henry, M. "The State of Water in Living Systems: from the Liquid to the Jellyfish." *Cellular and Molecular Biology* 51 (2005): 677–702.

Hodnett, R. M. "Water Storage Facilities and Suction Supplies." In *Fire Protection Handbook*. Quincy, MA: National Fire Protection Association, 1981.

Hofkes, E. H. and J. T. Visscher. "Renewable Energy Sources for Rural Water Supply." In *Technical Paper Series* 23. The Hague: IRC International Water and Sanitation Centre, 1986.

Horton, Robert, E. The Horton Papers. 1933.

Intergovernmental Panel on Climate Change. *Climate Change 2007: Impacts, Adaptation and Vulnerability. Contribution of Working Group II to the Fourth Assessment Report of the Intergovernmental Panel on Climate Change*. Cambridge, UK: Cambridge University Press, 2007.

International Union of Pure and Applied Chemistry. IUPAC 1997. Compendium of Chemical Terminology.

Jiangfeng, W. and P. A. Dirmeyer. "Dissecting Soil Moisture-Precipitation Coupling." *Geophysical Research Letters* 39, no. 19 (2012).

Johnson, D. L., S. H. Ambrose, T. J. Bassett, M. L. Bowen, D. E. Crummey, J. S. Isaacson, D. N. Johnson, P. Lamb, M. Saul, and A. E. Winter-Nelson.

"Meanings of Environmental Terms." *Journal of Environmental Quality* 26 (1997): 581–589.

Jones, A. A. *Global Hydrology: Processes, Resources, and Environmental Management.* England: Longman, 1997.

Jordan, T. D. *A Handbook of Gravity-Flow Water Systems for Small Communities.* London: Intermediate Technology Publications, 1984.

Kalenik, M. *Zaopatrzenie w wodę i odprowadzanie ścieków.* Warsaw: SGGW, 2009.

Kasting, J. and D. Catling. "Evolution of a Habitable Planet." *Annual Review of Astronomy and Astrophysics* 41, no. 1 (September 2003): 429–463.

Kennish, M. J. Practical Handbook of Marine Science. 3rd ed. Boca Raton, FL: CRC Press, 2001.

Konikow, L. and E. Kendy. "Groundwater Depletion: A Global Problem." *Hydrogeology* 13 (2005): 317–320.

Larson, E. E. and P. W. Birkland. *Putnam's Geology.* New York: Oxford University Press, 1994.

Lauterjung, H. and G. Schmidt. *Planning of Intake Structures.* Braunschweig: Vieweg, 1989.

Lautridou, J. P. and J. C. Ozouf. "Experimental Frost Shattering: 15 Years of Research at the Centre de Geomorphologie du CNRS." *Progress in Physical Geography* 6, no. 2 (1982): 215–232.

Lawrence, E., A. R. W. Jackson, and J. M. Jackson. "Eutrophication." In *Longman Dictionary of Environmental Science.* London: Addison Wesley Longman Limited, 1998.

Leclerc, H., L. Schwartzbrod, and E. Dei-Cas. "Microbial Agents Associated with Waterborne Diseases." *Critical Reviews in Microbiology* 28, no. 4 (2002): 371–409.

Ledford, H. "The Biological Higgs." *Nature* 483, no. 7391 (March 2012): 528.

Lerner, K. L. and B. Wilmoth Lerner. "Hydrosphere." In *The Gale Encyclopedia of Science*, Vol. 3. 4th ed. Detroit: Thomson Gale, 2008, 2213–2214.

Levy, Y. and J. N. Onuchic. "Water Mediation in Protein Folding and Molecular Recognition." *Annual Review of Biophysics and Biomolecular Structure* 35 (2006): 389–415.

Lomborg, B. The Skeptical Environmentalist: Measuring the Real State of the World. Cambridge, UK, and New York: Cambridge University Press, 2001.

L'vovich, M. I. *World Water Resources and Their Future.* Washington, DC: American Geophysical Union, 1979.

Mangelsdorf, J. *River Morphology: A Guide for Geoscientists and Engineers.* Berlin: Springer-Verlag, 1990.

Manzoor, K. P. "The Global Water Crisis: Issues and Solutions." *The IUP Journal of Infrastructure* 9, no. 2 (June 2011): 34–43.

Marshak, S. "A Surface Veneer: Sediments, Soils and Sedimentary Rocks." In *Earth Portrait of a Planet.* 3rd ed. New York: W.W Norton and Co., 2008.

Mathur, A. and D. da Cunha. *Mississippi Floods: Designing a Shifting Landscape.* New Haven, CT: Yale University Press, 2001.

Mattos, C. "Protein-Water Interactions in a Dynamic World." *Trends in Biochemical Sciences* 27, no. 4 (April 1, 2002): 203–208.

Meinzer, O. E. "Outline of Ground-water Hydrology with Definitions." *US Geological Survey Water-Supply Papers* 494 (1923).

Miller, S. L. "Production of Amino Acids under Possible Primitive Earth Conditions." *Science* 117 (1953): 528–529.

Molden, D. ed., International Water Management Institute, and Comprehensive Assessment of Water Management in Agriculture. Water for Food,

Water for Life: a Comprehensive Assessment of Water Management in Agriculture. London: Earthscan, 2007.

Morvan, H. P., S. Wardeh, N. G. Wright, "Desalinization for Syria." In *Food Security Under Water Scarcity in the Middle East: Problems and Solutions*, edited by A. Hamdy and R. Monti, 325–336. Options Méditerranéennes: Série A. Séminaires Méditerranéens 65. Bari: CIHEAM, 2005.

Moss, A. J. and P. Green. "Sand and Silt Grains: Predetermination of Their Formation and Properties by Microfractures in Quartz." *Australian Journal of Earth Sciences* 22, no. 4 (1975): 485–495.

Mulder, M. *Basic Principles of Membrane Technology.* Kluwer Academic Publisher, Dordecht, Western Netherlands, 1996.

Nahon, D. and R. Trompette. "Origin of Siltstones: Glacial Grinding Versus Weathering." *Sedimentology* 29 (1982): 25–35.

NASA. "NASA Probe Data Show Evidence of Liquid Water on Icy Europa." *U.S. Newswire*, November 16, 2011.

Nathan, C. C. *Corrosion Inhibitors*. Houston, TX: National Association of Corrosion, 1973.

Neuendorf, K. K. E., J. P. Mehl, and J. A. Jackson eds. *Glossary of Geology.* 5th ed. Alexandria, Virginia, American Geological Institute, 2005.

Oki, T. and S. Kanae. "Global Hydrological Cycles and World Water Resources." *Science* 313, no. 5790 (2006): 1068–1072.

Pal, S. K. and A. H. Zewail. "Dynamics of Water in Biological Recognition." *Chemical Reviews* 104, no. 4 (April 2004): 2099–2123.

Papavinasam, S. "Corrosion-Inhibitors." In *Uhlig's Corrosion Handbook*, 2nd ed., edited by R. Winston Revie, p. 1089–1105. New York: Wiley, 2000.

Parker, D. C. and R. Berry. "Clear Skies on Mars." *Astronomy* 21, no. 7 (July 1993): 72.

Paul, J. "Water Vapor in Atmosphere Key to Weather: Final Edition." *Milwaukee Journal Sentinel*, September 5, 1998, 14.

Poehls, D. J. and G. J. Smith. Encyclopedic Dictionary of Hydrogeology. Elsevier Academic Press, Amsterdam, The Netherlands, 2011, 528 pages.

Prinz, D. and S. Wolfer, "Traditional Techniques of Water Management to Cover Future Irrigation Water Demand." *Zeitschrift für Bewässerungswirtschaft* 34, no. 1 (1999): 41–60.

Punmia, B. C., A. Jain, and A. Jain. *Water Supply Engineering.* New Delhi: Laxmi Publications, 1995.

Radford, T. "Masses of Water Discovered in Space." *The Guardian London*, April 8, 1998. http://www.crystalinks.com/astronomy5.html

Rao, P. V. *Intermediate Vocational Course, First Year: Water Supply Engineering.* Hyderabad, Andhra Pradesh, India: Telugu Akademi, 2005.

Rasaiah, J. C., S. Garde, and G. Hummer. "Water in Nonpolar Confinement: From Nanotubes to Proteins and Beyond." *Annual Review of Physical Chemistry* 59 (2008): 713–740.

Ritter, M.E. "The Physical Environment: an Introduction to Physical Geography: The Geologic Work of Streams." Visited: March 2, 2008. http://www.uwsp.edu/geo/faculty/ritter/geog101/textbook/title_page.html

Rodríguez, S., C. Almquist, T. G. Lee, and M. Furuuchi. "A Mechanistic Model for Mercury Capture with In Situ-Generated Titania Particles: Role of Water Vapor." *Journal of the Air and Waste Management Association* 54, no. 2 (February 2004): 149.

Ron, C. "Earth, Water, and Comets." Science News 152, no. 7 (August 16, 1997): 107.

Rothschild L. J. and R. L. Mancinelli. "Life in Extreme Environments." *Nature* 409, no. 6823 (February 2001): 1092–1101.

Sata, T. *Ion Exchange Membranes: Preparation, Characterization, Modification and Application.* London: Royal Society of Chemistry, 2004.

Satterlund, D. R. and P.W. Adams. *Wildland Watershed Management.* 2nd ed. New York: John Wiley and Sons, Inc., 1992.

Sawyer, C. N. and P. L. McCarty. *Chemistry for Sanitary Engineers.* 2nd ed. New York: McGraw Hill, 1967.

Schroeder, B. "Industrial Water Use." Part of Water is Life, a class website (http://academic.evergreen.edu/g/grossmaz/VANOVEDR/) on water privatization and commodification produced by students of Geography 378 (International Environmental Problems and Policy) at the University of Wisconsin-Eau Claire, United States, spring 2004.

Shaheen, E. I. and W. Chantarasorn. "Transport of Shoal Deposits." *Journal of the Water Pollution Control Federation* 43, no. 5 (May 1971).

Shiklomanov, I. A. *World Water Resources: Modern Assessment and Outlook for the 21st Century.* St. Petersburg: Federal Service of Russia for Hydrometeorology and Environment Monitoring, State Hydrological Institute, 1999.

Singh Jurel, R., R. B. Singh, S. Kumar Jurel, R. Singh. "Infiltration Galleries: A Solution to Drinking Water Supply for Urban Areas near Rivers." *IOSR Journal of Mechanical and Civil Engineering* 5, no. 3 (January/February 2013): 29–33.

Stanley, S. L. Jr. "Amoebiasis." *The Lancet* 361, no. 9362 (March 22, 2003): 1025–1034.

Steele, R. G. "Universe is Full of Water." *Sarasota Herald Tribune.* 31/03/2001.

Stover, E. L., C. N. Haas, K. L. Rakness, and O. K. Scheible. *Design Manual: Municipal Wastewater Disinfection.* Cincinnati, Ohio: US Environmental Protection Agency, 1986.

Strahler, A. and A. Strahler. *Introducing Physical Geography.* Boston: Wiley and Sons, 2006.

Strathmann, H. "Electrodialysis." In *Membrane Handbook,* edited by W. S. W. Ho and K. K. Sirkar. New York: Van Nostrand Reinhold, 1992.

Strathmann, H. *Ion-Exchange Membrane Separation Processes.* New York: Elsevier, 2004.

Swarthout, D. and C. M. Hogan, 2010. "Stomata." *Encyclopedia of Earth.* National Council for Science and the Environment, Washington, DC.

Swilling, J. *Minerals: Key to Vibrant Health and Life Force.* Jacob Swilling Health and Fitness, p. 163, Lulu Press 2004.

Taylor, K., 2011. "Biggest Cloud of Water in Universe Discovered," TG DAILY. http://www.tgdaily.com/space-features/57445-biggest-cloud-of-water-in-universe-discovered

Trevors, J. T. and G. H. Pollack. "Hypothesis: the Origin of Life in a Hydrogel Environment," *Progress in Biophysics and Molecular Biology* 89 (2005): 1–8.

Tuena de Gómez-Puyou, M. and A. Gómez-Puyou. "Enzymes in Low Water Systems." *Critical Reviews in Biochemistry and Molecular Biology* 33, no. 1 (1998): 53–89.

United Nations. "International Decade for Action: 'Water for Life' 2005–2015." UN-Water, 2007. http://www.un.org/waterforlifedecade/scarcity.shtml

United Nations. "World Population Prospects: The 2012 Revision, Key Findings and Advance Tables." Working Paper No. ESA/P/WP.227. New York: UN Department of Economic and Social Affairs, Population Division, 2013.

United Nations. "World Water Development Report." New York: United Nations, 2003.

United Nations Environment Programme. *Marine Pollution from Land-Based Sources: Facts and Figures*. Paris: UNEP Industry and Environment, 1992.

United Nations Environment Programme. "Vital Water Graphics: An Overview of the State of the World's Fresh and Marine Waters." Retrieved 2002. http://www.unep.org/dewa/assessments/ecosystems/water/vitalwater/

United Nations Educational, Scientific and Cultural Organization. "Water for People, Water for Life: United Nations World Water Development Report." UNESCO, 2003. p. 36.

United Nations Educational, Scientific and Cultural Organization. "Water: a Shared Responsibility." The United Nations World Water Development Report 2. 22/03/2006.

Untersteiner, N., 1975. "Sea Ice and Ice Sheets and Their Role in Climatic Variations." *The Physical Basis of Climate and Climatic Modelling*, Global Atmospheric Research Project (GARP) Publication Series 16, World Meteorological Organization/International Council of Scientific Unions, pp.206–224.

USSR Committee for the International Hydrological Decade. *World Water Balance and Water Resources of the Earth*. Paris: UNESCO, 1978.

Vaclavik, V. A. and E. W. Christian. *Essentials of Food Science*. New York: Springer, 2007.

Valenti, M. "Power Where Sun Meets Sea." *Mechanical Engineering* 118, no. 2 (February 1996): 154.

Vymazal, J. "Constructed Wetlands for Wastewater Treatment." *Water* 2, no. 3 (2010): 530–549. doi:10.3390/w2030530.

Water Pollution Control Federation Disinfection Committee. *Wastewater Disinfection: A State-of-the-Art Report*. Alexandria, VA: Water Pollution Control Federation, 1984.

Wayman, E. "How Did Early Earth Have Liquid Water?" *Science News*. April 29, 2013.

Welle, K., J. Tucker, A. Nicol, B. Evans. "Is the Water Sector Lagging behind Education and Health on Aid Effectiveness? Lessons from Bangladesh, Ethiopia and Uganda. *Water Alternatives* (2009) 2 (3): 297–314.

White, G. C. *Handbook of Chlorination*. New York: Van Nostrand Reinhold Company, 1978.

Wilson, W. E. and Moore, J. E. *Glossary of Hydrology*, Alexandria, VA: American Geological Institute, Springer, 2003.

World Glacier Monitoring Service. *Fluctuations of Glaciers 1990–1995*. Vol. VII. Zurich, Switzerland: IAHS (ICSI)/UNEP/UNESCO, 1998.

World Health Organization. Water Sanitation and Health (WSH). "Facts and figures: Water, sanitation and hygiene links to health." 2004. http://www.who.int/water_sanitation_health/publications/factsfigures04/en/

World Meteorological Organization. *Comprehensive Assessment of the Freshwater Resources of the World*, Geneva: WMO, 1997.

World Resources Institute, United Nations Environment Programme, United Nations Development Programme, and World Bank. *A Guide to the Global Environment*. New York: Oxford University Press, 1998.

Zhong, D. "Hydration Dynamics and Coupled Water-Protein Fluctuations Probed by Intrinsic Tryptophan." *Advances in Chemical Physics* 143 (July 2009): 83–149.

Zhong, D., S. K. Pal, and A. H. Zewail. "Biological Water: A Critique." *Chemical Physics Letters* 503, no. 1 (2011): 1–11.

Zronik, J. P., *Jacques Cousteau: Conserving Underwater Worlds*. Crabtree Publishing Company, New York, 2007.

CPSIA information can be obtained
at www.ICGtesting.com
Printed in the USA
LVIC06n2043271114
415880LV00014B/43

* 9 7 8 1 4 9 4 7 1 8 9 9 2 *